NUMBER WONDERS

NUMBER WONDERS

171 Activities to Meet Math Standards & Inspire Students

BY CATHERINE JONES KUHNS

Crystal Springs BOOKS

SDE

a division of Staff Development for Educators

Peterborough, New Hampshire

Published by Crystal Springs Books
A division of Staff Development for Educators (SDE)
10 Sharon Road, PO Box 500
Peterborough, NH 03458
1-800-321-0401
www.sde.com/crystalsprings

Published 2006
Printed in the United States of America
12 11 10 09   2 3 4 5

ISBN: 978-1-884548-91-8

Library of Congress Cataloging-in-Publication Data

Kuhns, Catherine Jones, 1952-

 Number wonders : 171 activities to meet math standards & inspire
   students / by Catherine Jones Kuhns.

     p. cm.

 Includes bibliographical references.

 ISBN: 978-1-884548-91-8

 1. Mathematics—Study and teaching—Activity programs—United States.

 2. Mathematics—Study and teaching--Standards—United States. I.

Title. QA13.K84 2006

 372.7—dc22

                          2006015154

Editor: Sharon Smith
Art Director, Designer, and Production Coordinator: Soosen Dunholter
Photographer: Catherine Jones Kuhns
Illustrator: Soosen Dunholter

To my great husband, George, and my wonderful children, Adam and Meaghan, and in memory of my parents, Marianne and Jim Jones, my first and best teachers

CONTENTS

ACKNOWLEDGMENTS

I am blessed to teach at Country Hills Elementary with the most innovative and caring group of teachers I can imagine. These professionals put children first and inspire me each day I work with them.

Thanks to Carol Newman for encouraging me to think outside the box and to Donna Morrison for allowing me to do so.

Thanks also to Char Forsten, who opened new doors for me; to Sharon Smith, my patient and meticulous editor; to Soosen Dunholter, whose graphics gave this book life; and to Lorraine Walker and Amy Aiello, who gave me the final push to get this book together.

INTRODUCTION

In 2000, the National Council of Teachers of Mathematics (NCTM) released *Principles and Standards for School Mathematics*. Since then, these guidelines have served as models for state and district math standards and for expectations for each grade band.

Specifically, NCTM has identified 5 content strands and 5 processing strands for mathematics, and has established standards and expectations for each of those strands by grade band: pre-kindergarten through grade 2, and grades 3–5, 6–8, and 9–12. This book focuses on the content strands: Number and Operations, Algebra, Geometry, Measurement, and Data Analysis and Probability. My goal is to provide meaningful lessons and activities for students in kindergarten through grade 2, addressing every one of the NCTM content standards and expectations.

But good math instruction does more than just cover the content. Research has shown that for a math lesson to be successful (for students to understand and retain it), it should not just involve important mathematical concepts. It should begin with a problem that is interesting, and it should connect to prior knowledge (Hiebert 1997). The goal of excellent mathematics instruction is to create engaging lessons in which the students enjoy using mathematics, deepen their understanding of math, connect their learning to the outside world, and learn new skills. The lessons in this book serve as a way to reach that level of joy, understanding, connection, and mastery.

How do they do that? You'll notice that the activities in this book include a great deal of mathematical dialogue and writing, not only for the teacher but also for the students. If they're to become good mathematicians, children need to be able to speak and write mathematically. However, they will be able to do that only when they have heard mathematics spoken correctly, have spoken mathematically themselves, have seen mathematical writing, and have been a part of the writing process.

I'm a big believer in math journals (some people call them math logs). Math journals work well as one place to keep track of mathematical thinking. Writing helps the writer make sense of things. If you give children simple spiral notebooks, even kindergarteners can "record" in picture form what they learn. Always encourage drawings, sketches, and diagrams (Whitin 2000).

As the year goes on, your students will become better at recording what they've learned—and that is part of what makes a math journal such a wonderful way to show growth over time at a parent conference. Imagine an April parent conference at which you show the parents how their child wrote about math in September compared to how that child is writing about math in April! That is pure authentic assessment.

Students need to see you model mathematical writing before they work in their math journals. They also need to hear you *speak* correct mathematical language. Students should hear math spoken using the correct terms. Children can use geometric terms just as easily as they can rattle off the names of dinosaurs and game-card characters. Kids *love* to use big words! It makes them feel smart.

Today many intermediate teachers find themselves needing to reteach math vocabulary because their students were never told in primary school that a square is also a rectangle, a rhombus, and a polygon. That's so unnecessary! And it won't be an issue later on if you treat your little problem-solvers as mathematicians. Say "vertical," not "up and down." Study the math glossary in the back of your textbook and make sure you're using the correct terms. Your students will rise to the expectations you set.

It's also important to give children time to practice talking mathematically. Don't expect your math instructional time to be quiet. When children talk with one another, they are helping each other make sense of what they're learning. Their discussions with one another help "cement" their understanding. Rosemary Irons, a noted expert in helping young children understand math, has said, "Do not let them go to centers by themselves. They won't talk!"

When you want your class to talk mathematically, tell them that. Say, "Boys and girls, I need you to discuss this with your friends using math talk." That lets your students know that you're expecting to hear only math conversations, not talk about Little League, ballet class, soccer practice, or TV shows. Students need to know that this also means they're to talk softly so they don't disturb other mathematicians and their conversations.

When your class is engaged in math discussion, praise those students who use correct terminology. Children need to hear their teacher say things like, "Great math language, Jimmy. I like how you called it a square, not a box." Or "Wonderful, Linda. I like how you called that the *x*-axis." Since children want to please their teacher, they will claim those terms for themselves and in turn start to communicate mathematically.

In short: Grab every opportunity to model for your students the correct way to write and speak mathematically. At first you'll need to be deliberate in your conversations. You'll be thinking, "Gee, is there a way I can add a little math talk here?" But after a little practice, I promise, you too will become a better mathematics communicator! Use the lessons and activities in this book to give children plenty of practice. Set high expectations. And recognize it when students meet those expectations. That's how you start to build strong mathematicians.

And now, let the mathematics begin!

A Quick Tip

In some classes it may be necessary to demonstrate what "math talk" sounds like and looks like so everyone understands what's expected. Don't be surprised if you have to demonstrate "math talk" more than once.

NCTM STANDARDS & EXPECTATIONS

● Number & Operations Standard for Grades Pre-K–2

	Expectations	Page
Instructional programs from prekindergarten through grade 12 should enable all students to—	**In prekindergarten through grade 2 all students should—**	
Understand numbers, ways of representing numbers, relationships among numbers, and number systems	• count with understanding and recognize "how many" in sets of objects;	20
	• use multiple models to develop initial understandings of place value and the base-ten number system;	28
	• develop understanding of the relative position and magnitude of whole numbers and of ordinal and cardinal numbers and their connections;	32
	• develop a sense of whole numbers and represent and use them in flexible ways, including relating, composing, and decomposing numbers;	35
	• connect number words and numerals to the quantities they represent, using various physical models and representations;	38
	• understand and represent commonly used fractions, such as ¼, ⅓, and ½.	40
Understand meanings of operations and how they relate to one another	• understand various meanings of addition and subtraction of whole numbers and the relationship between the two operations;	42
	• understand the effects of adding and subtracting whole numbers;	44
	• understand situations that entail multiplication and division, such as equal groupings of objects and sharing equally.	45
Compute fluently and make reasonable estimates	• develop and use strategies for whole-number computations, with a focus on addition and subtraction;	49
	• develop fluency with basic number combinations for addition and subtraction;	52
	• use a variety of methods and tools to compute, including objects, mental computation, estimation, paper and pencil, and calculators.	53

Algebra Standard for Grades Pre-K–2

	Expectations	Page
Instructional programs from prekindergarten through grade 12 should enable all students to—	**In prekindergarten through grade 2 all students should—**	
Understand patterns, relations, and functions	• sort, classify, and order objects by size, number, and other properties;	61
	• recognize, describe, and extend patterns such as sequences of sounds and shapes or simple numeric patterns and translate from one representation to another;	64
	• analyze how both repeating and growing patterns are generated.	67
Represent and analyze mathematical situations and structures using algebraic symbols	• illustrate general principles and properties of operations, such as commutativity, using specific numbers;	70
	• use concrete, pictorial, and verbal representations to develop an understanding of invented and conventional symbolic notations.	71
Use mathematical models to represent and understand quantitative relationships	• model situations that involve the addition and subtraction of whole numbers, using objects, pictures, and symbols.	73
Analyze change in various contexts	• describe qualitative change, such as a student's growing taller;	74
	• describe quantitative change, such as a student's growing two inches in one year.	75

Geometry Standard for Grades Pre-K–2

	Expectations	Page
Instructional programs from prekindergarten through grade 12 should enable all students to—	**In prekindergarten through grade 2 all students should—**	
Analyze characteristics and properties of two- and three-dimensional geometric shapes and develop mathematical arguments about geometric relationships	• recognize, name, build, draw, compare, and sort two- and three-dimensional shapes; • describe attributes and parts of two- and three-dimensional shapes; • investigate and predict the results of putting together and taking apart two- and three-dimensional shapes.	79 82 84
Specify locations and describe spatial relationships using coordinate geometry and other representational systems	• describe, name, and interpret relative positions in space and apply ideas about relative position; • describe, name, and interpret direction and distance in navigating space and apply ideas about direction and distance; • find and name locations with simple relationships such as "near to" and in coordinate systems such as maps.	90 91 92
Apply transformations and use symmetry to analyze mathematical situations	• recognize and apply slides, flips, and turns; • recognize and create shapes that have symmetry.	93 95
Use visualization, spatial reasoning, and geometric modeling to solve problems	• create mental images of geometric shapes using spatial memory and spatial visualization; • recognize and represent shapes from different perspectives; • relate ideas in geometry to ideas in number and measurement; • recognize geometric shapes and structures in the environment and specify their location.	99 100 104 105

Measurement Standard for Grades Pre-K–2

Instructional programs from prekindergarten through grade 12 should enable all students to—	Expectations In prekindergarten through grade 2 all students should—	Page
Understand measurable attributes of objects and the units, systems, and processes of measurement	• recognize the attributes of length, volume, weight, area, and time;	109
	• compare and order objects according to these attributes;	112
	• understand how to measure using nonstandard and standard units;	115
	• select an appropriate unit and tool for the attribute being measured.	116
Apply appropriate techniques, tools, and formulas to determine measurements	• measure with multiple copies of units of the same size, such as paper clips laid end to end;	117
	• use repetition of a single unit to measure something larger than the unit, for instance, measuring the length of a room with a single meterstick;	118
	• use tools to measure;	119
	• develop common referents for measures to make comparisons and estimates.	120

Data Analysis & Probability Standard for Grades Pre-K–2

Instructional programs from prekindergarten through grade 12 should enable all students to—	Expectations In prekindergarten through grade 2 all students should–	Page
Formulate questions that can be addressed with data and collect, organize, and display relevant data to answer them	• pose questions and gather data about themselves and their surroundings; • sort and classify objects according to their attributes and organize data about the objects; • represent data using concrete objects, pictures, and graphs.	127 129 130
Select and use appropriate statistical methods to analyze data	• describe parts of the data and the set of data as a whole to determine what the data show.	131
Develop and evaluate inferences and predictions that are based on data	• discuss events related to students' experiences as likely or unlikely.	137
Understand and apply basic concepts of probability	(There are no specific expectations given for this standard.)	139

NUMBER & OPERATIONS

Before students can begin to deal with higher-level mathematical concepts, they need to develop number sense—that basic understanding of what numbers are and how they relate to each other. Number sense is so much more than computation skills. It's knowing when to add, when to subtract, when to multiply, and when to divide. Number sense also means literally a "sense of numbers"—an understanding of numbers so complete that a child knows that 6 is half of 12 and that it's also 3 doubled, 1/3 of 18, 2 sets of 3 or 3 sets of 2, 1 more than 5 and 1 less than 7, and that if you add 10 to it, you'll get 16. Now *that* is real number sense!

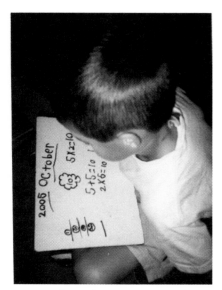

In October, the tenth month of the year, ask your little math detectives how many number sentences they can find for the number 10.

When it comes to creating number sense, counting is a fundamental building block (Copley 1999). Young children should count, count, count! They need to see and hear you counting as well. Count chairs, glue sticks, steps to the cafeteria, steps to the water fountain, rungs on the monkey bars, fingers, girls, boys, anything. Just count!

And then move to the next step. Children need to be able to do more than recite the numbers. They need to connect the numbers to the counting order (Van de Walle 2003). They need to see you touching objects as you count them and they need to see you counting from left to right. Then they need to practice those things themselves. This helps children learn a systematic way of counting and avoid counting the same thing twice. It's also important to demonstrate that counting in a different order does not change the total number of objects (NCTM 2000).

One way to establish strong number sense is through the use of manipulatives. Why use manipulatives? Jean Piaget, the Swiss psychologist, observed that children must experience the same concept in a variety of ways in order to truly understand that concept (Ginsburg 1998). Just because a child can count 6 cookies does not mean he can count to 6. It means he can count 6 cookies! That child must count 6 cookies, 6 chips, 6 plastic frogs, 6 pennies, and more to really understand 6. So pull out the manipulatives, counters, or whatever you call them, and let the kids *count!*

And once they've conquered counting, don't put those manipulatives away. For centuries, research has been confirming the effectiveness of manipulatives in teaching multiple concepts in mathematics.

And that's why manipulatives figure so prominently in this chapter.

STANDARD Understand numbers, ways of representing numbers, relationships among numbers, and number systems.

EXPECTATION: Count with understanding and recognize "how many" in sets of objects.

 ## Count, Count, Count!

Don't wait until "Math Time" to count—seize every opportunity! When you call your students to reading group, say, "I have 3 panthers from the panther group at the reading table. I should have 6. I'm missing 3 panthers." When they've arrived, add, "Great, 3 more panthers came to reading group. Now I have all 6 panthers!"

When you walk around your class, say, "Let's see. There are 5 children at this table. I like how all 5 are on task."

As students line up for playground, say, "I have 3 kids in my line. Oh, now I have 3 more; that's 6. Great, 2 more have joined us. Let's see: how many are in my line now?"

In first and second grades, children should hear, "Okay, I need 6 kids for reading. I have 2 ready with their books. That means two-sixths of my group is ready with four-sixths of the group still missing." Or "Four-fifths of this table of 5 children are ready for writing."

Use Manipulatives

Jean Piaget theorized that children must construct knowledge from the concrete in order to move on to formal operations or abstract tasks. He advocated child-centered classrooms in which children could make discoveries through hands-on learning. Piaget believed that active involvement fostered learning (Wadsworth 1989). Those of us who use manipulatives in our classrooms today know that Piaget was right.

I strongly recommend that you use a variety of manipulatives—and use manipulatives mats, too. Manipulatives mats help each child keep his manipulatives in one place. To create mats, start with the patterns on pages 146-55. Cut out the individual pattern pieces and trace around the pieces on appropriate colors of construction paper. Cut out the colored pieces and glue them into place on other sheets of construction paper. If you laminate the completed mats, you can use them over and over again for years.

RULES FOR MATHEMATICIANS & THEIR TOOLS

- Only lucky and responsible mathematicians get to use these math tools.

- Use the math tools only in the way the teacher demonstrates.

- Never throw or chew a math tool or put it in your pocket or backpack.

- Take what you need and keep those tools on a work mat or at your table. Always put the math tools back exactly where you found them.

Store manipulatives in clear containers, and it's easy for students to find what they need.

That is, you can use them for years if they're taken care of. I've found it's important to place the manipulatives and the mats in a spot where the children can access them easily. Demonstrate how you want them cared for and what your expectations are. Set ground rules and stick to them!

My manipulatives are all within easy reach. Clear containers let students see the contents. Small bowls and baskets rest on top, so

a child who doesn't want or need to take all the manipulatives from one container can pull out just some of them.

When the kids are learning how to use the manipulatives and the mats, I like them to sit close to me. If the students are on the floor and I'm standing above them, I can easily see if they're using the manipulatives correctly. The following activities represent some of the ways my students learn with manipulatives.

With the kids on the floor, the teacher sees more—so it's a breeze to assess their use of manipulatives.

Shirt Manipulatives Mat

Make a copy of the reproducible on page 146. Use that as a pattern to create a construction-paper mat for each student. (Alternatively, you can skip the tracing step if you simply copy the reproducible onto colored stock.) Give each child a "shirt," along with a handful of pom-poms to use as buttons. Instruct students to put 3 pom-poms on the left side and 4 on the right side—thus practicing directionality and at the same time proving that 3 + 4 = 7.

Using the mat makes it clear: 3 pom-poms + 4 pom-poms = 7 pom-poms.

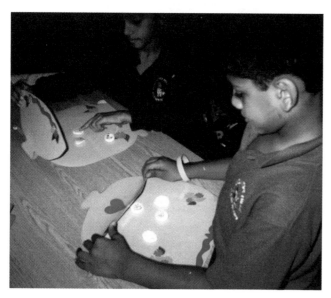

Always begin with the concrete. Here, 3 cookies + 1 cookie = 4 cookies.

Cookie-Jar Manipulatives Mat

Make a copy of the reproducible on page 147. Use that as a pattern to create a construction-paper mat for each student. (Alternatively, you can skip the tracing step if you simply copy the reproducible onto colored stock.) After you cut out each tracing, fold it on the dotted line to simulate the lid of a cookie jar. Give each student a "cookie jar" plus pieces of cereal to represent cookies. Students may decorate their cookie jars with stickers or flowers. Have each child put 5 "cookies" on the jar, then lift the lid and slide 2 cookies under the lid. Hey, 5 − 2 = 3! Or you can have each student pull 3 cookies out from under the lid and add them to the 2 cookies she's already placed on the jar. She's just demonstrated that 3 + 2 = 5.

Ocean Manipulatives Mat

Start by copying the patterns on page 148, cutting out the copies, and tracing them onto appropriate colors of construction paper. Each child will need a blue "wave" to represent the ocean; aside from that, you can cut out as many or as few fish and pieces of coral as you want from whatever colors are available.

Create manipulatives mats by gluing each wave onto a full sheet of construction paper to represent the ocean. Let kids decorate their mats with the fish and pieces of coral. (If you prefer, you can copy the fish and coral onto plain white paper and let the children color and cut them out.) Then have them practice math facts by using gummy worms, gummy sharks, plastic boats, seashells, or paper sea creatures as manipulatives.

Students enjoy practicing counting skills with miniature plastic boats.

Pond Manipulatives Mat

Create a mat for each child by cutting a free-form pond shape from blue construction paper. Copy the flower and leaf patterns on page 149, cut out the pieces, and trace them onto appropriate colors of construction paper. Then cut apart the construction-paper pieces and position them appropriately around the edges of each "pond." Hand out Goldfish crackers or plastic fish, ducks, frogs, or boats as manipulatives.

Introduce the mats with instructions something like these: "Boys and girls, listen closely to my story. Move your Goldfish crackers to match my words. Ready? There were 6 fish in the pond. A hungry bear walked by and—*swoop!*—he grabbed 3 fish. How many fish were left?"

Or you might say, "There were 7 fish in the pond. Did you each place 7 fish in your pond? Then 1 more fish jumped from the river into the pond. How many fish are in the pond now?" Other addition sentences can be demonstrated with stories like this: "There were 5 fish swimming in the right side of the pond and 3 fish swimming in the left side of the pond. How many fish were in the pond?"

To follow up, invite students to practice the math facts the class is working on by creating and telling their own math-pond stories. A child might say, "There were 4 fish in the pond. Then Todd went fishing and he caught 2 fish. How many are left?"

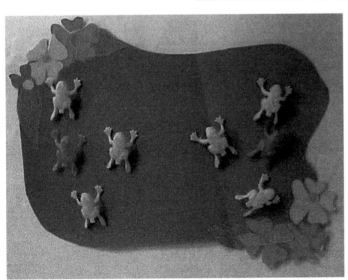

"There were 7 frogs in the pond, and then 1 more jumped in. How many frogs are in the pond now?"

Manipulatives aren't just toys. They're "gifts"!

MANIPULATIVES ARE MORE THAN TOYS!

Do parents question your use of "toys" or "games" when they really mean manipulatives? Do they suggest that this is a new, trendy idea in education? The use of manipulatives is not new. In fact, this practice has been around a long time.

Johann Pestalozzi, a Swiss educator who taught disadvantaged youth in the late eighteenth and early nineteenth centuries, observed that students needed to learn through physical activity. They should work with objects before words, he said, and with the concrete before the abstract (Resnick 1998).

Influenced by Pestalozzi, Germany's Friedrich Froebel, the Father of Kindergarten, filled nineteenth-century classrooms with what he referred to as "gifts." There were 20 specific gifts—objects such as balls, hoops, and sticks—placed in each class.

Later, in the U.S., a young Frank Lloyd Wright attended a kindergarten that replicated Froebel's approach. This was the same Frank Lloyd Wright who went on to become an architectural genius. He credited the gifts as the foundation of his interest in and understanding of architecture. Now *that's* powerful!

"Put 3 sprinkles on the vanilla scoop and 4 on the strawberry scoop. How many sprinkles do you have in all?"

● Ice-Cream-Cone Manipulatives Mat

Start with the patterns on page 150. Make a copy and cut out the individual pieces. Trace 1 ice-cream cone for each student onto brown paper. Cut out the tracings. Repeat the process with the scoops of ice cream, creating 1 pink and 1 white scoop per child. For each mat, glue 1 cone, 1 pink scoop of ice cream, and 1 white scoop onto a large sheet of construction paper, creating a double-scoop cone. This color combination lets you say, "Put 6 sprinkles on the strawberry scoop and 4 sprinkles on the vanilla." Students see that 6 + 4 = 10. Use small candy pieces, bingo chips, or pom-poms for sprinkles.

● Flower-Garden Manipulatives Mat

For this mat, you're going to make a simple flower garden with flowerless stems. The number of stems is determined by the number you're working on. For each pair of students, you'll need 1 sheet of blue construction paper (9 x 12 inches for numbers 3 through 11 or 12 x 18 inches for numbers over 12), half a sheet of green construction paper the same width as the blue paper, odds and ends of construction paper in black and yellow, and 2 groups of counters, each group a different color. You'll also need green, red, and black markers.

To make each mat, trace the grass pattern on page 151 onto the green construction paper and cut out the "grass." (If necessary, extend the pattern to match the longer dimension of the blue construction paper.) Position the blue paper horizontally and glue the grass in place at the bottom, matching the illustration. Next, starting where the grass leaves off, draw a series of straight green lines for stems. You may want to color in a few leaves as well. Make the number of stems equal the number that you want your children to practice. If you're working on 17, draw 17 stems.

Here's real flower power! Children use counters to show all the possible combinations that add up to your target number.

Trace the pattern for the butterfly wings onto yellow construction paper and the body onto black construction paper; to dress things up, glue the butterfly onto the mat. I like to add a bright red smile to the butterfly's head, and to draw in a couple of ladybugs (or use stickers) as well, just to make things interesting and appealing to students.

In class, pair off students and give each pair a flower-garden mat and a small pile of counters in each of the colors you've chosen. Instruct children to use the counters to create flowers. At the same time, explain that they should manipulate the counters to find all possible number sentences for 17. If they use 12 red counters (12 red flowers) and 5 white counters (5 white flowers), then 12 + 5 = 17. They can also show 17 as 6 red and 11 white flowers, 10 red and 7 white, and so on.

● Bedtime Manipulatives Mat

Start with a mat of white poster board cut in half widthwise. Set the mat in front of you horizontally and draw a line across the sheet 7 inches from the top edge. Divide the rest of the mat (the bottom section) into 12 equally sized spaces. This is the bear's quilt. Copy and then cut out the reproducible patterns on page 152. Trace the pieces of the teddy bear and his pillow onto construction paper, cut out the construction-paper pieces, and glue them in place on the poster board to create the heading shown in the illustration.

I have 6 of these mats for my class. The mats take time to create, so they are "special" and can be found only at a center. As always, it's important to make sure students understand how to use these mats before turning anyone loose with them. Have one child demonstrate how to use the mat. Provide her with manipulatives (any manipulative that can fit inside one of the spaces on the mat) in 3 different colors. Instruct the child to place one manipulative on each of the spaces. If the bed has 4 of one color, 2 of another, and 6 of a third, then 4 + 2 + 6 = 12!

This bedtime mat takes a little more prep time, but it's worth it.

Volcano Manipulatives Mat

For this mat you'll need construction paper in black, orange, green, brown, gray, and blue, plus red tissue paper.

Make copies of the reproducible patterns on pages 153–55 and cut out the individual pieces. For each mat, trace the mountain onto black construction paper, the large lava flow onto orange construction paper, and the small lava flow onto red tissue paper. (Using the construction paper and the tissue paper creates a neat effect.) Use gray construction paper for the smoke. Repeat with green construction paper and the grass pattern from page 151. Cut out all of the pieces and glue them onto a sheet of brown construction paper. Add a strip of blue at the top for the sky, making the mat look something like the illustration.

The tissue paper creates a special effect, and plastic dinosaurs are ideal manipulatives.

In class, hand out the mats. For manipulatives, put out gummy worms or plastic dinosaurs, skeletons, or insects. These will quickly become favorite mats and manipulatives for any and all dinosaur lovers.

To begin the lesson, say, "Okay, class, I need you to move your plastic dinosaurs to match my math stories." Continue, "Long ago in the land of dinosaurs, 4 dinos were at the watering hole when 3 more dinosaurs joined them. How many dinosaurs were slurping up water?" Or you might say, "10 dinosaurs were near the mountain when 7 of them heard a noise that frightened them. All 7 ran off to hide. How many creatures were left at the mountain?" Stories like this guarantee that everyone will be listening and practicing math facts.

A QUICK TIP

It's always easier to *say* an answer than it is to *explain* an answer. Explaining an answer requires much more thinking and gives your students a chance to speak mathematically. Their conversations will be a window into your students' understanding or misunderstanding of a concept. If you ask a question that can be answered in one word, that right answer can be a lucky guess, rather than a true indication of understanding!

 ## S t r e t c h a Counting Book!

Collect a variety of counting books. Look for counting books that start at 1 and increase and for those that start at 10 and decrease to zero. Also collect number books that go beyond 10.

When reading a counting book, ask children to predict what the number will be on each page. When you read the page with 6 kangaroos say, "6 kangaroos. How many tails? How many ears? How many eyes? How many back feet? How many front feet? If each kangaroo has 1 joey, how many joeys?" Say, "Before I turn this page, what number of animals do you predict will be on the next page?" Make it a bit more challenging by saying, "What makes you think that?"

Practice "Counting On"

Scenario: You ask a child what's 1 more than 4 or what's 4 plus 1. He looks at you with big eyes and starts counting on his chubby little fingers, saying, "1, 2, 3, 4 . . . *[long pause]* 1." Then he starts the whole process all over again until he either gets it or just shrugs and quits.

What does that tell you about the child?

It means that the child doesn't really understand how many things that first number represents. So he needs to go back and count 4 many, many, many times. He should count 4 cubes, 4 paper clips, 4 cereal pieces, 4 blocks, and so on until he's comfortable with 4.

The Counting-On Jar

A clear plastic jar and a few Unifix cubes help kids with counting on.

One strategy to help students over the hump of "counting on" uses a clear plastic peanut butter or parmesan cheese jar (empty and clean, of course) and Unifix cubes. The strategy goes something like this:

- Have the children sit close to you.

- Say, "Let's put 3 Unifix cubes in this jar. Can you help me?"

- As a child drops cubes in the jar, you and the class count out loud: "1, 2, 3."

- Screw the top onto the jar.

- Shake the jar and say, "There are 3 cubes in this jar, right?"

- Students agree.

- Ask, "Can we 'count on' using this jar? Watch. We don't have to count these 3 cubes because we know there are 3 cubes in here, right? So, what's 1 more than 3?"

- Shake that jar once, making a *loud* noise as you say, "3."

- Without shaking the jar again, say, "4!"

- Toss the jar (it won't break) to a child and ask, "What's 1 more than 3?"

- The child shakes the jar 1 shake (no doubt making a louder noise than you did) and says, "3, 4!"

- Tell the shaker to toss the jar to a friend.

- With a new child holding the jar, ask, "Now, what is *2* more than 3?"

- The child shakes the jar and says, "3, 4, 5!"

Doing this for numbers 3 through 10 will help children learn to count on.

Palming

Once students no longer need the plastic jar used in the previous activity, invite them to "palm" the number for counting on.

Let's say you want students to count on in order to add 6 + 2. Teach them first to decide which number is greater, 6 or 2. Model the process by flattening out one hand and making a fist with the other hand. As you use your fist to hit your palm, say the number that is greater—in this case, 6. Then count with 2 fingers: "7, 8."

This method works well for counting on or for adding 1, 2, or 3.

 ## Letters in Names

Read *Tikki Tikki Tembo* by Arlene Mosel and *Chrysanthemum* by Kevin Henkes to your class. Because the names of the title characters in these books have many letters, they are perfect lead-ins to activities that deal with the numbers of letters in the children's names. Say something like, "Gee, Chrysanthemum has 13 letters in her name. How about you? How many letters are in *your* name?" Then have each child build a tower of Unifix cubes, using one cube for each letter in his first name. With older children, have each student build *2* towers, one for his first name and one for his last.

To encourage discussion, model by saying something like, "My name is Catherine. I have 9 letters in my name and 9 cubes in my tower. My tower is taller than Leslie's but shorter than Christopher's. Leslie's name has 3 fewer letters than my name. Christopher's name has 2 more letters than mine. Now turn to your buddies and talk about your first-name towers. Use lots of math talk. Compare the numbers of cubes." For older children, repeat with last names and with first and last names combined into one tower.

EXPECTATION: Use multiple models to develop initial understandings of place value and the base-ten number system.

 ## Bean Sticks

A bean stick is a craft stick that has 10 lima beans, garbanzo beans, or other beans of a similar size glued to it. Bean sticks make great manipulatives because they are proportional; the beans fill the stick, so the stick is 10 times the size of 1 bean.

Working from the concrete to the symbolic helps students develop an understanding of proportionality that's critical for developing place-value knowledge (Van de Walle 2003). This proportional model accomplishes that.

These tried-and-true sticks are a popular way to demonstrate and help students understand the base-ten system.

Begin by having each child make 10 bean sticks. (Do not be tempted to ask a parent or aide to do this for your students. The children need to do it themselves so that they *really believe* there are 10 beans on each stick.)

Once the bean sticks are made, continue the lesson over the course of many days. Don't rush it. Some days you will accomplish only one or two steps. That's okay. Consistency pays off! The students *will* get it. Teaching place value just takes time— and constant reinforcement.

Here's how to work with the bean sticks:

- Give 10 craft sticks to each child.

- The child writes her name on one side of each stick and then turns the stick over.

- Using a paper plate as a work mat, the child pours a good amount of glue from one end of a stick to the other. (This is the one time you don't have to say, "Not too much glue, please.")

- She places 10 beans on the stick, and then repeats the process for each of her other sticks.

- Leave the paper plates overnight so the glue dries completely.

- The next day, say, "There are 10 beans on each stick. We don't have to count the beans since you know you already counted them, right?"

- Children should agree.

- Say, "Show me 10 beans."

- Each child should pull out 1 stick.

- Say, "Show me 20 beans."

- Each child should pull out 2 sticks.

- Continue to do this until children are sure that each stick represents 10.

- Next, give each child a cup containing 10 individual beans.

- Say, "Show me 12 beans."

- Children should show 1 stick and 2 loose beans.

- Ask why this is a more efficient or better way to show 12 than just showing 12 loose beans.

- Continue this process for 24, 54, and 35, reinforcing the concept, for example, that 35 is 3 sticks and 5 loose beans.

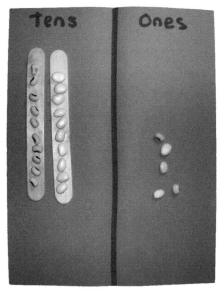

These mats are easy for kids to make, and they're great for practicing 2-digit numbers.

Tens-and-Ones Mats

This is a way to take the bean sticks to the next level.

- Give a sheet of construction paper to each child.

- Demonstrate folding the paper in half vertically to crease it, then unfolding it. Children should follow your lead.

- With crayon or marker draw a line along the fold. Children should do the same.

- On the top left side, demonstrate how each child should write "Tens."

- On the top right side, demonstrate how each child should write "Ones."

- Say, "When I ask you to show 25, please put 2 bean sticks on the tens side and 5 loose beans on the ones side."

- Practice this *many* times with many different numbers.

Moving On with Bean Sticks

Once students have mastered the basic concept of tens and ones, it's time to build on that knowledge. Once again, make sure every student has bean sticks, loose beans, and his tens-and-ones mat. Then proceed along these lines:

- Say, "Show 64."

- Make sure everyone has done this correctly, placing 6 bean sticks on the left side of the mat and 4 loose beans on the right.

  - Say, "Add 10 more."

  - Each child should add 1 more bean stick.

  - Say, "Let's count. Point to your bean sticks and beans. Start with the tens."

  - With you leading, students should count, "10, 20, 30, 40, 50, 60, 70, 71, 72, 73, 74."

  - As they gain understanding, ask children to add 20, 30, and so on.

  - Once this is accomplished, ask children to add 24, 32, and so on.

  - Be prepared to see amazing things out of your students.

ADDING ZERO

Our base-ten system makes perfect sense. It's predictable, it has a pattern, and there are really only 10 numerals, 0 through 9, to learn. We didn't always have this system. It came to the Western world about a thousand years ago from the Hindus. They got it from the Arabs. The only thing the Hindus added was zero—a very important addition!

Chains of 10

Another way to reinforce the crucial benchmark of 10 is with chains of 10. The chains can be constructed from paper, paper clips, or plastic links. Each day of school, add a link to the chain. Once you have a chain that's 10 links long, hang it where the class can see it; then begin another chain. For example, on Day 57 of school, the children will see 5 separate complete chains and another chain with only 7 links. Ask the class to predict how long the chain will be on 100 Day when you connect all the chains.

Number Line

Use adding-machine tape or colored sentence strips to create a number line to hang in the room. Add a number each day. Be sure to change to a different color of marker or sentence strip each time you start a new set of 10 so that, once again, the decades stand out. Each time the count reaches 10, 20, 30, 40, and so on, highlight that number with glitter, a smile in the 0, or a star. (If you prefer, you can create a number line from cutouts that are consistent in shape and color for each group of ten—e.g., numbers 1–10 on red triangles, numbers 11–20 on pink circles, numbers 21–30 on orange squares.)

To truly celebrate the days with zeros, students can take turns wearing a cape you've made and being "Zero the Hero" for the day!

Who is that caped wonder?
Could it be "Zero the Hero"?

Pennies to Dimes

Our monetary system is rooted in base ten. Reinforce that real-world connection by adding (or having a student add) a penny to a clear plastic sandwich bag each day. On the tenth day trade 10 pennies for a dime. Place the dime in a separate clear sandwich bag next to the pennies bag. Count the money each day. On day 45, the children should count, "10¢, 20¢, 30¢, 40¢, 41¢, 42¢, 43¢, 44¢, 45¢!"

Standards are listed with the permission of the National Council of Teachers of Mathematics (NCTM). NCTM does not endorse the content or validity of these alignments.

31

EXPECTATION: Develop understanding of the relative position and magnitude of whole numbers and of ordinal and cardinal numbers and their connections.

● Stand in Order

Use and reuse this simple activity to work in content instead of fluff during those few minutes at the end of the day.

You'll need 10 large cards. Before class, number the cards (1 number per card) from 1 through 10. In class, proceed something like this:

- Give 1 card to each of 10 students.

- Ask them to arrange themselves in front of the class in number order, but add, "You may not talk!" Can they do this? Watch to see who can and who needs guidance.

- Once they get themselves in order, say, "If I call your number, raise your card above your head. Ready?"

- Say, "1, 3, 5, 7."

- Watch as they raise their cards. (Some will need reminders of what numbers they're holding.)

- Now ask the whole class: "What do you notice about who has a number raised?"

- Accept answers and discuss.

- "Who *doesn't* have a number raised? Why?"

- Accept answers and discuss.

- Continue slowly, saying, "All numbers down. Now, if you are number 5, step forward."

- Wait.

- Say, "If you are 1 *more than* 5, step forward. If you are 1 *less than* 5, step back."

- Continue, "Okay, everybody back to your original place in the line. Let's do another one. If you are number 4, step forward. If you are *more than 4,* step forward. If you are *less than 4,* step back."

- "Now go back to your line. Thank you."

- Try another one: "If you are even, step forward 2 steps."

- Continue with further commands to reinforce 1 more, 1 less, 2 more, 2 less, odd and even, greater than, and less than.

- Ask students not holding cards to come up with more commands. They love the opportunity to be the boss!

Variations for older students: Start with cards marked with the even numbers from 122 to 144. Ask the card-holding students to stand in order. Say, "Will the number that's 16 more than 124 please step forward." Or "If you're holding the number that is 1 hundred, 3 sets of ten, and 6 ones, please raise your card." If your class is really in need of a challenge, include mixed numbers, so students must arrange themselves as, say, 1, 1½, 2, 2½, 3, 3½, 4, 4½, 5.

Ordinal Memory Game

This is a version of the Memory Game (a.k.a. Concentration). Reserving at least 4 students to make up the audience for this activity, cut out enough pairs of duplicate magazine pictures so that each student who's not in the audience can have 1 picture. Mount each individual picture (not each pair) on a separate sheet of black paper. Then, in class, proceed as follows:

- Pass out the pictures, being careful to make sure no one sees anyone else's picture. Hand out no more than 1 picture per child. Some children will not receive any pictures; those students are the audience.

- Each child with a picture holds it close, with the back of the paper facing the audience so that no one can see the picture. These children stand at the front of the room and count off, saying, "I'm first," "I'm second," and so on.

- Audience members take turns predicting which pictures will match. If an audience member says, "I'd like to see third and sixth," then the third child and the sixth child reveal their pictures. If the pictures match, then those students hand their pictures to the teacher before sitting down with the audience.

- If the pictures do *not* match, the third child and the sixth child turn their papers so the pictures are once again hidden. They stay in the line.

- Each time a pair of students joins the audience, the students who remain standing renumber themselves, calling out their new positions in order.

- Each time they renumber themselves, ask questions like, "Why did the names after *fifth* change?" Or if the first child left, ask, "Why did everyone's name change?"

- The game continues until all pictures have been matched.

Find It!

For each child, you'll need 1 clear disk and 1 copy of the reproducible 100 Chart Game Board on page 156. The first few times you do this, it helps to have a game board on the overhead as well. That way you can move a chip on the overhead as the children move the chips on their game boards.

Go *very slowly* through the steps. This is a good 10- to 15-minute activity to supplement a lesson. Children will need lots of repeated practice (maybe over several days) with each step, and some steps may need to be broken into more than one lesson or even spread out over more than one week. The process goes like this:

- Hand out the game boards and chips.

- Say, "Place your disk on 25." This alone is a task for the young child; model it on your overhead.

- Say, "After you place your disk on 25, add 1, and then add 1 again. Where are you?"

- Give other, similar directions starting at different numbers.

- Say, "Place your disk on 45, and add 10." This is your time for a little drama as you move your chip on the overhead 10 times. Say, "Where are you? *Wow!* There has to be an easier way to move this many spaces! What did you notice happened when I moved 10?"

- You may have to guide students by pointing out where your chip was and where it is now. You hope someone will see that adding 10 means dropping the chip to the space directly below where it was.

- Say, "Look! The number in the ones place in 45 is 5, and the number in the ones place in 55 is also a 5. Why do you think that is?" Encourage discussion and observations. Point out the similar patterns throughout the chart.

- Continue by saying, "What do you notice about the numbers in the tens place each time we add 10?" It is important that the children see that each time 10 is added, the number in the ones place stays the same and the number in the tens place increases by 1.

100 Chart Game Board

1	2	3	4	5	6	7	8	9	10
11	12	13	14	15	16	17	18	19	20
21	22	23	24	25	26	27	28	29	30
31	32	33	34	35	36	37	38	39	40
41	42	43	44	45	46	47	48	49	50
51	52	53	54	55	56	57	58	59	60
61	62	63	64	65	66	67	68	69	70
71	72	73	74	75	76	77	78	79	80
81	82	83	84	85	86	87	88	89	90
91	92	93	94	95	96	97	98	99	100

"Place your disk on 25 on the game board. Add 1, and then add 1 again."

THE 100 CHART GAME BOARD

Some people call it a 100 Chart. I like to call it a game board. Whatever you call it, why is it so important? When they work with this game board/chart, children are studying the patterns in our number system without realizing it. They become pattern seekers. This helps children in your class and later on in their mathematics education. The more your students work with the game board, the more the patterns become "engraved" in their brains, aiding them in mental computation.

- Say, "Okay, so whenever we want to add 10, we drop to the number below. Let's practice that. Start at 67 and add 10."

- Ask, "Then how would I add 20?" Discuss with the class.

- "Place your disk on 63. Add 10, add 5, add 20. Where are you?"

- "Hands behind your back. Start at 28, add 10, add 10, add 2. Where are you?"

- Ask children to come up with other directions.

- Follow a similar routine with subtraction.

INSTRUCTIONS FOR MATH JOURNALS

"Write 4 steps or clues using your 100 Chart Game Board. Tell the starting number, give 4 steps to move the chip, and then give the ending number."

EXPECTATION: Develop a sense of whole numbers and represent and use them in flexible ways, including relating, composing, and decomposing numbers.

 ## Dose of Daily Digits

My own particular version of calendar strategies is a routine I call my Dose of Daily Digits. During Daily Digits children are problem-solving, using the numbers in the date and in the count of how many days they've been in school.

NCTM recommends that teachers "ensure that students repeatedly encounter situations in which the same numbers appear in different contexts" (NCTM 2000). To come up with a different number to work with each day, I use the number of days we've been in school that year, the number of days we've been in school that month, and/or the date. Switching things around ensures that, over time, we have opportunities to work with the same numbers in many different situations.

My calendar set-up is very eclectic. I use elements of Box It, Bag It; Math Their Way; and Every Day Counts; plus my own twists. You can mount pennies on your bulletin board in clear envelopes or snack bags, and add 3 containers for straws representing ones, tens, and hundreds. (My straw containers sit on the ledge at the bottom of my calendar board.) I use boxes of graduated size, but just about any containers—including French-fry containers—will work. You'll also need to post plain paper for recording tally marks each day; that paper stays up all year.

Let's say it's the 38th day of school. On your bulletin board you already have 37 tally marks left from yesterday, so you just add 1 more to show 38. Then say something like, "Gee, how many sets of 5 are there in 38?" Point to each set of 5 as you count. "Let's see. There are 1, 2, 3, 4, 5, 6, 7 sets of 5 and [counting again] 1, 2, 3 left over." Remember, we want children to know that 38 is not just 1 more than 37 or 1 less than 39; it is also 7 sets of 5 with 3 left over. That's number sense!

"This is the 38th day of school. How many sets of 5 are there in 38?"

Either you or a student should move straws to show 38 as 3 sets of 10 in a tens box and 8 loose straws representing ones.

Use Unifix cubes to represent each school day in the month, adding 1 more cube for each school day. Place the cubes on a shelf or counter so that the children can easily see them. If you keep the cubes in the same place all day, they will be handy reminders of the day's number. You may want to label the cubes with a sign that reads something like, "This is the 12th day we've been in school this month."

Snap the cubes together in sets of 2. Count the cubes by 2s. If it's the 12th day of school this month, slide the cubes apart with 4 to the left and 8 to the right and ask, "What number sentence does this show for 12?" Keep separating the cubes in different combinations to allow children to see that 12 is 4 + 8, 6 + 6, and 10 + 2. Don't forget 4 + 4 + 4 or 2 + 4 + 6!

That works well for half of the days, but the other half will be odd numbers. Seize this as a perfect opportunity to reinforce odd numbers. On the 9th day the children will see 9 as 4 sets of 2 with 1 left over. Say, "Let's count by 2s. Ready?" Point and say, "2, 4, 6, 8" and then whisper "9." Now, loud and clear, add, "Look at this. There are 9 cubes. 4 sets of 2 and 1 cube that is not in a set of 2. What does that tell us about 9?" Elicit discussion. "That's correct: 9 is odd and I know that because 1 of the cubes does not belong to a set of 2."

The problems that I give use all 4 number operations. The class works with models and word problems to understand how to use these operations. "Today is 10/14/06. We have been in school 38 days. Today we can play with these digits: 10, 14, 6, and 38."

If you're working with first or second graders, you could then say, "There are 10 boxes and each box has 14 cookies. How many cookies are there in all?" Write "10 x 14 = ." As you write, say, "10 sets of 14." Then continue, "Let's pull this apart and see if it makes it easier. Is 14 the same as 10 and 4? Then let's pull 14 apart. Do you know 10 sets of 10? Right! 10 sets of 10 is 100. Put that in your brain and remember that. Now, remember how we pulled 14 apart to 10 and 4? So what is 4 sets of 10? Right: 40. So, add the 100 in your brain to 40. Yes, 140!

"So if 10 boxes with 14 cookies in each box is 140, you should be able to solve this. Luis made 14 pizzas and he put 10 pepperoni slices on each pizza. How many pepperoni slices did he need? Yes, 140." Write "14 x 10 = 140."

Continue: "Remember that 10 sets of 14 is 140. That will help you solve this. There were 140 goldfish divided evenly among 14 bowls. How many goldfish to a bowl?" Write "140 ÷ 14 = ." Say, "Yes, there are 10 goldfish to a bowl."

"Now, can you solve this? There were 140 Little Leaguers divided equally among 10 baseball fields. How many players to a field?" Write "140 ÷ 10 = ." Say, "Yes, there would be 14 to a field." Fill in the 14.

Finally, you might say, "There are 180 days in the school year. We've been in school 38 days. How many days are left?" Write "180 − 38." Either you or a student can count out Popsicle sticks to *show* students the problem. For older students, show the algorithm.

KINDERGARTEN QUESTIONS FOR THE DOSE OF DAILY DIGITS

When working with kindergarteners or other early mathematicians, keep your numbers simple. Most likely, with kindergarteners you would not use 14 on October 14. You might use 10 for the date and then 2 just because it is easier for them to understand. (You can take this date thing only so far!) Some classes will need to work with numbers like 5 or 6 for months. Of course, you can still work separately with the 38 that represents the 38 days you've been in school. Choosing from those options, ask your kindergarteners questions like these:

- "There are 10 cookies in the cookie jar. I add 2 more. How many are there now?" On the board or on chart paper, write the number sentence "10 + 2 = 12."

- "There are 2 pennies in my bank. I add 10 pennies to the bank. How many pennies are in the bank now?" Write "2 + 10 = 12."

- "I had 12 crayons and I lost 2. How many crayons do I have now?" Write "12 − 2 = 10."

- "There were 12 bluebirds in the tree but 10 flew away. How many are in the tree?" Write "12 − 10 = 2."

- "Would you rather have 38 dollars or 180?" As the kids answer, write "38 < 180" and "180 > 38."

- "Could you hold 38 grains of sand in your hand? 38 nickels? 38 grapes?"

- "Could we fit 38 bunnies in this class? 38 children? 38 elephants? 38 school buses?"

- "How many pennies are in 38 cents? How many dimes? Nickels?"

- "I'm so very hungry. Could I buy a lunch for 38¢?"

- "Could I buy a meal with $38? A refrigerator? A car?"

- "Stand up. Close your eyes. Sit down when you think 38 seconds have passed."

- "Are we in school more or less than 38 minutes?"

- "What might last 38 minutes?" If your class is ready, show this on a manipulative clock.

EXPECTATION: Connect number words and numerals to the quantities they represent, using various physical models and representations.

Number-Line Cards

If you create number-line cards with your students instead of using the store-bought number line, you'll increase the number of ways your students use the number line.

To create your own class number-line cards, you'll need one 12-inch square of black paper for each number. In addition, you'll need a 13-inch square of red paper for each odd number and a 13-inch square of blue paper for each even number.

For younger students: Each day you'll create *one* card with your students. The first day you do this, mount a black square on top of a red square. Hang this square where all the children can see it and where you can reach it. Then proceed along these lines:

- Say, "We're going to start our own number cards today. We'll make one card a day. Today we'll start with 1."

  - Add, "Watch as I make a 1."

  - Using chalk, draw a 1 in the corner of a black square.

  - Say, "I'm going to glue 1 thing in the middle."

  - Glue 1 large item—or a large picture of 1 item—in the center.

  - Say, "1 is odd. So I'm going to write that on the square. Watch as I write this: 'O-n-e is odd.'"

  - That's it for day 1. After the children leave, hang that card where you usually hang your number line.

  - The next day, show the class a black square mounted on a blue square. Say, "Today we will make the card for 2. Watch as I write '2' in the corner."

  - Write a 2 in the corner.

  - "Now we need to glue 2 feathers in the center." Ask a child to glue the feathers in the center.

  - Say, "2 is an even number, so watch as I write this. 'T-w-o is even.' Notice that we've glued number card 2 onto a different color than what we used for 1. Why do you think we did that?"

Give each odd number a red "frame."

Each day continue the red-blue-red-blue pattern of cards, and continue gluing objects in place to show each number. You can use balloons, googly eyes, sequins, or other fun things from the craft store. You can use different objects every day, but of course all the objects used for a particular number need to be the same. If you're working with kindergarteners, continuing this strategy through 10 or 12 will provide your students with great exposure to basic math concepts.

For older students: Each day, create *one* card with your students. As with the younger children, each day you'll use a 12-inch black square mounted on a 13-inch red or blue square. The first day you create the cards, mount a black square on a red square. Hang this square where all the children can see it. Then proceed along these lines:

- Say, "We're going to start making our own number cards today. We'll make one card a day. Today we'll start with 1. Watch as I make a 1."

- Draw a 1 in the corner.

- "I'm going to glue 1 thing in the middle." Glue a large picture or a large item in the center.

- Say, "1 is odd. So I'm going to write that. Watch as I write this: 'O-n-e is odd.' Also, mathematicians have another special word for one. It is *unique*. That means that one is unlike any other number. So watch as I write this: '1 is unique.'"

- At the end of the day, place this card where it will stay for the rest of the school year.

- On the second day, pull out a black square mounted on blue. Say, "Today we have a card for 2. I'll make a 2 in the corner."

- Continue, "This square is on blue, and yesterday's square was on red. Why do you think I've done this?" Elicit ideas.

- Say, "Those are great predictions. We'll have to wait until we have more cards up to see if you're correct."

- "Now today I have some feathers. How many arrays can I create for 2?" Depending upon when you are doing this and the grade level, you may need to define and demonstrate arrays.

- Survey the class for answers to your question about arrays. "Yes, you're correct. I can arrange 2 feathers in 1 vertical row or 1 horizontal row."

- Ask students to help you glue 4 feathers into the 2 arrays (1 vertical and 1 horizontal). Use chalk to circle each horizontal row of each array.

- Say, "This array is 1 horizontal line, so I'm going to write below it, '1 set of 2.'" As you say this, write "1 x 2." The word "times" means nothing to young children; it's confusing. But "1 set of 2" makes sense to them.

- Turn to the vertical line and say, "This array is 2 sets of 1, so watch as I write that. '2 sets of 1 .'" Then write "2 x 1."

- Continue this day after day until you get into the teens and reach the number you desire for your class.

- Let's say you've reached the 12th day. You would say, "As you can see, the black square is mounted on blue because 12 is even." Write, "12 is even."

- "Now watch as I write 12 and be thinking of all the arrays that I can arrange for 12."

- You may need to write the arrays on the board or chart as the children generate a list.

- When all possible arrays have been named, plan with the class how to arrange the arrays and glue them in place.

- Circle each horizontal row of the array that shows 4 rows of 3 and say, "This is 4 sets of 3." Write "4 x 3."

- Add, "Since there are more than 2 arrays for 12, we know 12 is composite." Write "12 is composite." on the bottom of the square.

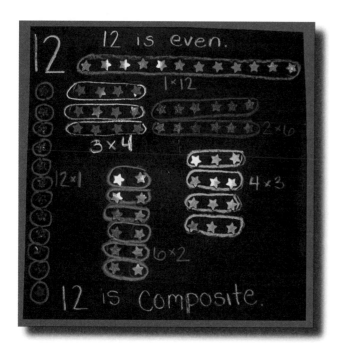

You'll need to use smaller items with numbers like 12, 16, and 18 since there are so many arrays that can be made for those numbers. Small sequins, beans, stars, or beads work well.

EXPECTATION: Understand and represent commonly used fractions, such as ¼, ¹/₃, and ½.

 Name Fractions

Have second graders write math fraction sentences about their names. One child might write: "My name is Adam. My name is ²/₄ or ½ vowels. My name is ²/₄ or ½ consonants. My name is ½ A, ¼ D, and ¼ M." Another child would write, "My name is Meaghan. My name is ³/₇ vowels and ⁴/₇ consonants. My name is ¹/₇ M, ¹/₇ E, ²/₇ A, ¹/₇ G, ¹/₇ H, and ¹/₇ N."

Family Fractions

First and second graders can write simple fraction sentences about their families. A child might write, "My family is $1/4$ Dad, $1/4$ Mom, and $2/4$ kids!" Another might write, "My pets are $1/3$ cats and $2/3$ fish!"

Animal Fractions for First & Second Grades

Ask each student to draw 8 different animals. Each student's animals should include at least 1 bird, fish, mammal, reptile, insect, and amphibian. Once all drawings are complete, have each child write at least 6 sentences about her pictures. Those sentences might include something like, "$1/8$ of these animals hops. $3/8$ of the animals do not have legs. $1/8$ of the animals lay eggs."

Half-Time Day

At the halfway point of the school year, celebrate the fraction one-half. Call it Half-Time Day. Ask kids to come to school in sports-team shirts or cheerleading uniforms. You wear a referee's shirt and carry a whistle. Every half hour, blow the whistle and do something silly, like run in place, for 30 seconds.

Begin the day with a batch of cookies. Take a cookie recipe and cut it in half with the help of the children. Then bake exactly the right number of *large* cookies to equal half the number of students in your class. (They'll wonder, "What about the other half?" You know they'll break the cookies in half to share at the end of the day.)

Read Bruce McMillan's *Eating Fractions* or Stuart Murphy's *Give Me Half!* and stop at the halfway point. Finish the story later that day.

Ask kids to run 50 yards while you time them. Then ask them to run 25 yards and time them again. Compare times. They'll be very surprised.

HALF-TIME CENTERS

For Half-Time Day, set up centers where each student can:

- Draw half her face.

- Count objects and decide how many are half of the total.

- Stamp fruit or vegetables that have been cut in half and dipped in paint.

- Make one peanut-butter-and-jelly sandwich with a friend. Cut it in half, then cut the halves in half, the fourths in half, and so on. (Unless your class has future Wolfgang Pucks, they will not get past sixteenths.) The conclusion is for the friends to cut their *one* napkin in half and then for each to eat her half of the sandwich. (Check for peanut allergies before setting up this center.)

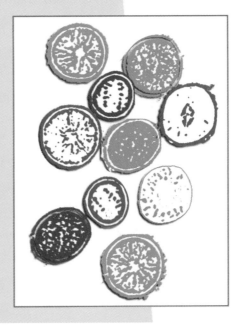

EXPECTATION: Understand various meanings of addition and subtraction of whole numbers and the relationship between the two operations.

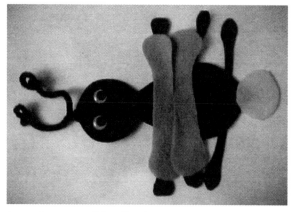

Turning into "fireflies" gives students another way to internalize math facts.

Flashing Fireflies

Ten Flashing Fireflies, by Philemon Sturges, is a great book for helping children understand the crucial benchmark of 10. The children see an addition number sentence for 10 on one page and on the next page they see the inverse (subtraction) number sentence for 10. They see that $9 + 1 = 10$ and $10 - 1 = 9$. Pure genius, Philemon!

Before class, use the reproducible patterns on page 157 to make 10 "fireflies" out of yellow, black and gray felt or construction paper. Glue each firefly to a clothespin.

In class, read this story to the students once for pure enjoyment. After reading ask, "Can anyone find the math in this story?" Allow discussion time. Then proceed as follows:

- Create a "bug jar" outline on the floor using markers on paper, or using chalk (which vacuums up quickly) or yarn right on the floor. The jar outline must be large enough for 10 children to stand inside it.

- Select 10 students to be "fireflies."

- Clip a firefly clothespin to each of the 10 children.

- Ask the 10 child fireflies to flit about the classroom as you reread the story. Each time the children in the book capture a firefly, you capture a child and walk that child to the jar, where he must stay. Use this as an opportunity to count with the children how many fireflies are now inside the jar and how many are now outside, thus reinforcing the number sentences.

- Once all fireflies are in the jar, let them out one by one. Count down with each departing firefly.

- Read the book aloud again. This time, write the corresponding number sentence on a chalkboard or chart as you catch each firefly. For instance, you might say, "There are 4 fireflies in my jar and I'm adding 1 more, so watch as I write this: '$4 + 1 = 5$.'"

- The next time you read the story, create number sentences for subtraction. Work with the number of fireflies *in the sky* as that number decreases. As you capture your child fireflies, write *those* corresponding number sentences on the chalkboard or chart. For example, "There were 6 fireflies in the sky and I caught 1, so watch how I write this: '6 – 1 = 5.' There are 5 fireflies left in the night sky."

Later the book and the firefly manipulatives can become part of a reading or math center in which students retell the story.

Number Sentences in Your Name

This one is for children who can spell their first and last names.

- Before class, turn to the reproducible grid on page 158 and make a copy of it for each child.

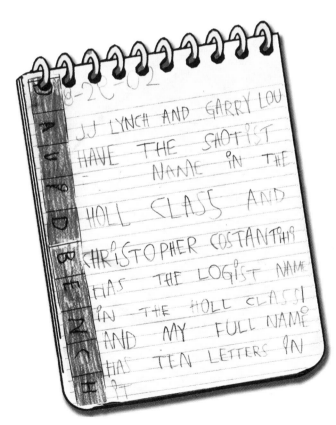

- In class, hand out the grids. Each child writes her first and last name on it in pencil, one letter to a square.

- Each student colors the squares with her first name in one color and the squares with her last name in another color.

- Each child cuts out her first and last names in one long strip. (Children with longer names may need to glue the first and last names together.)

- Say, "Each of you has a number sentence in your name. Look at Jinky's. Her first name has 5 letters. Her second name has 8 letters. Her number sentence is 5 + 8 = 13."

- Say, "We're going to create a chart of our name-number sentences from the least total to the greatest total."

- Ask, "Is there anyone in the class with a number sentence total of 1? Why not? How about 2?" Emphasize that no one has a total of 1 since it would be impossible to add 2 names and get 1.

- Keep calling out numbers until the child with the fewest letters says, "That's me!"

- Glue their number-sentence names onto chart paper in order of fewest number of letters to greatest number of letters.

This activity also provides great fodder for those math journals.

- Next to each name strip write the appropriate number sentence.

- When there are several number sentences that equal the same number, take advantage of this opportunity. Say, "Look at this: We have 4 names that equal 9. Each number sentence is different. Ted's name equals 9 and his number sentence is 3 + 6. Marc's number sentence is 4 + 5. Linda and Emily both have the number sentence of 5 + 4. But they each *equal* 9!"

- Keep this posted in your room for a while and refer to the number sentences.

EXPECTATION: Understand the effects of adding and subtracting whole numbers.

● Roll to One Hundred

This is a fun game that can be played for 5 to 15 minutes in groups of 2 to 5 children. Each child uses his 100 Chart Game Board and a marker. The group shares one die.

- Each child starts with a marker on number 1.

- One child rolls the die. If the roll is 6 then the child advances 6 numbers on his game board.

- All members of the group watch their friend to be sure he moves the correct number of spaces.

- Play goes to the next child, who rolls the die and moves that number of spaces.

- The winner is the child who reaches 100 first *or* the one who is closest to 100 when the teacher announces, "Time's up!"

Variation for children who can add: Each child in turn rolls 2 dice, totals the numbers on the dice, and moves that many spaces.

As children race through the 100 Chart Game Board, they're actually practicing addition facts.

● Roll to One

This game is similar to Roll to 100, but in this case each child starts with her marker on 100. Each roll of the die indicates how many spaces she needs to move *back*. The winner is the child who reaches 1 first or who is the closest to 1 when the teacher says, "Time's up!"

Variation for children who can add: Each child in turn rolls 2 dice, totals the numbers rolled, and moves back that number of spaces.

EXPECTATION: Understand situations that entail multiplication and division, such as equal groupings of objects and sharing equally.

● Number-Pattern Banners

Our number system is based on patterns. When children understand the patterns and begin to look for them, they become better, quicker, more confident "computers." They also see the relationships between numbers. This understanding deepens their number sense. Making Number-Pattern Banners helps children recognize and predict patterns in our number system. Working with a different banner each month reinforces skip counting, repeat addition, and multiplication.

Before hanging these banners on a bulletin board or from the ceiling, line up the "kid created creatures" (e.g., the cats, spiders, and so on from the following activities) on the floor. Put 1 creature in row 1, 2 in row 2, 3 in row 3, and so on, until you've used all pieces of artwork. If the number of students in your class results in a few extra creatures, use them as a border or as part of the title for your bulletin board. Then ask the children to help you find efficient ways to count the parts. If you were working with the Cats Banner, you'd say, "Let's count the cats' ears in row 4. Ready? 1, 2, 3, 4, 5, 6, 7, 8.

"Okay. Since there are 2 ears on each cat, let's count the same row by 2s. Ready? 2, 4, 6, 8.

"Is there another way to count this? Yes! 2 + 2 + 2 + 2 = 8!"

"How many cats do we have? How many ears? How many tails? How many paws?"

If your class is ready for this, add, "There are 4 cats with 2 ears each. That is 4 sets of 2 and that equals 8. That's 4 times 2 is 8 or 4 x 2 = 8." This interchangeable language helps everyone develop an understanding of the process.

Hmm. Patterns? This sounds a lot like algebra to me. You know that patterns are a major component of algebra so, yes, this would also be considered algebraic thinking.

● Cats Banner

There are many delightful stories with cats (think *Six-Dinner Sid* by Inga Moore, *The Tale of Tom Kitten* by Beatrix Potter, and *The Cat in the Hat* by Dr. Seuss). Read one of them with your students as a springboard for this banner.

Next, have each child draw a cat. Line up the cat drawings as described above. Then count together the number of paws, the number of ears, and the number of tails.

Your students are seeing patterns here. Since each row has 1 more cat than the row before, it is a growing pattern! And it's not just the number of cats that's increasing.

Standards are listed with the permission of the National Council of Teachers of Mathematics (NCTM). NCTM does not endorse the content or validity of these alignments.

45

The number of ears is increasing by 2s and the number of paws is increasing by 4s.

ART TIP: If you want the cats to be somewhat uniform in size, but you don't want to use patterns, tell each child to paint or draw a cat the size of her hand.

Spiders Banner

Before class, cut narrow strips of black paper for the spiders' legs, as it's difficult for kids to create these. (A paper cutter makes quick work of this.)

In class, read one of the Anansi tales. Anansi is a clever little spider who came to America with African storytellers. While Anansi and his tales are still popular in Jamaica, in the U.S. most tales that originated with Anansi have become Brer Rabbit stories. Gerald McDermott's *Anansi the Spider* is a wonderful story and his illustrations are great for inspiring children's artwork.

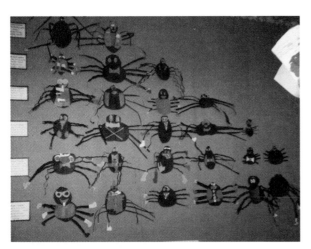
Count spider body parts by 2s and legs by 8s.

After reading the story, have children create freehand drawings of spiders. Put out black construction paper for the spider, scraps of colorful paper for the clothing, and the strips you already cut for legs. Show students how to fold and crinkle the strips for the legs. When all drawings are complete, put them together to form a growing pattern. In this pattern the number of body parts increases by 2s and the number of legs by 8s.

ART TIP: You want the spiders to be a uniform size, so tell children that each spider's body should be the size of the palm of the young artist's hand. To encourage colorful clothing suggest, "Make your spiders as funky and colorful as Gerald McDermott makes his."

Turkeys Banner

Student-created turkeys make a great banner when you place them in a growing pattern. Have students count the number of beaks by 1s, feet by 2s, and feathers by 4s.

Variation: Take the standard November turkey-hands art activity and kick it up a few notches. Cut a 6 x 9-inch piece of paper for each child. Put a variety of paint colors and brushes at the art center. Each child paints the palm of one hand brown and then uses a different color to paint each finger and the thumb of that hand. The student quickly presses his hand onto the paper, and then immediately turns to the sink for a rinse.

In November, count turkey beaks by 1s, feet by 2s, and feathers by 4s.

Reindeer Banner

To demonstrate making a reindeer, begin with a rectangular piece of brown construction paper. (I use the 12 x 18-inch size.) Position the paper horizontally. Form a cone, as in the illustration. Staple. Ask kids to follow your lead and then to trace each of their hands on a separate sheet of brown construction paper. Each child should cut out his tracings and glue or staple those to the top front of his reindeer for antlers. Allow kids to cut out eyes and noses for their reindeer from other colors of construction paper. More detail-minded children may add eyelashes.

Once the reindeer are created, start counting! Count noses by 1s. Count eyes by 2s. Count antler points, and you're counting by 10s.

In December, of course you count reindeer! And then you count eyes, noses, and antler points, too.

REINDEER COUNTING

You can vary the counting depending on the skills of your students. Alternatives might go something like this:

Basic	Advanced	More Advanced
4 reindeer	4 reindeer	4 reindeer
4 noses	1 + 1 + 1 + 1 = 4	4 x 1 = 4
8 eyes	2 + 2 + 2 + 2 = 8	4 x 2 = 8
40 antler points	10 + 10 + 10 + 10 = 40	4 x 10 = 40

Insects Banner

One of the focuses of insect study for primary students is learning that each insect has 3 body parts, 2 wings, 2 antennae, and 6 legs. Students follow that rule to create the insects on this banner. Then they count insects by 1s, wings or antennae by 2s, body parts by 3s, and legs by 6s.

ART TIP: Demonstrate folding paper and making cuts so that insects are symmetrical. (That would be geometry.) Demonstrate cutting the folded paper so that one cut results in 2 legs. To make the insects more uniform in size, set out a paper plate and tell the children to make each insect cover most of the plate.

Show students how to make colorful insects like this one. Then put their creations in yet another number-pattern banner—and start counting!

For Valentine's Day, count heart-shaped flowers and leaves—and reinforce the concept of symmetry at the same time.

 Valentine Flowers Banner

When children create Valentine flowers, each flower has 2 heart-shaped leaves and 4 heart-shaped petals. Line up the pieces of artwork. When children count flowers, they're counting by 1s. When they count leaves, they're counting by 2s, and when they count petals, they're counting by 4s.

ART TIP: Demonstrate how to cut a symmetrical heart shape by folding a piece of paper and cutting half of a heart shape, then opening up the folded paper. Hand out sheets of paper approximately 6 x 9 inches on which the students can mount their heart flowers. For something different, provide pipe cleaners for stems.

● **Cookie Division**

This activity is based on the book *The Doorbell Rang,* by Pat Hutchins. Begin by reading the story to the class for enjoyment. Then proceed along these lines:

● Place a small bowl of Cookie Crisp cereal at each table (or substitute any other cereal pieces that could represent cookies). Tell kids to count out 12 "cookies."

● As you reread the story, have students divide their supplies of cookies to show how the children in the story share the cookies with the friends who come into the kitchen.

● Ask each child to write a cookie problem in which children have to figure out how to divide cookies.

ART TIP: Stick a 3-inch-wide strip of black-and-white check or "wood grain" contact paper to the bottom of a page a student will draw on. Repeat for each student. This strip becomes the kitchen floor, like the one in *The Doorbell Rang,* for each child's story illustration. Invite children to draw characters to match their cookie stories.

Stick a 3-inch-wide strip of checkered or "wood-grain" contact paper to the bottom of a page a student will draw on. This is the kitchen floor.

The Doorbell Rang

Writing Math Stories

Great word-problem *solvers* are also great word-problem *writers*. Often it is harder to *write* math word problems than it is to solve them. I ask my students to write 3 problems during the course of a week. On Friday, each child selects his favorite one for me to edit—but before he brings the problem to me, he reads it to at least 2 classmates, and the classmates try to solve the problem. A common concern in writing word problems is leaving out important information; classmates will let the author know if information is missing.

Many times the child gets to take that word problem to the publishing stage, typing it on the computer and adding artwork. This step builds confidence, but don't let it become the main focus. What's important is the process of *developing* the problem, not the actual typing and decorating. Younger children will need the teacher to print or type their stories.

STANDARD Compute fluently and make reasonable estimates.

EXPECTATION: Develop and use strategies for whole-number computations, with a focus on addition and subtraction.

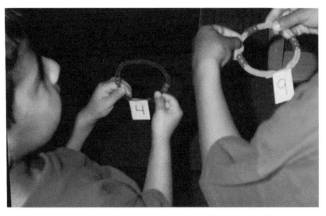

I have everyone use the same color pipe cleaner for the number we're working on, and then use a different color for the next number. That way I can say, "Get out your red 8 bracelets. Let's find number sentences for 8!"

Number Bracelet

Give each student a pipe cleaner (all the pipe cleaners should be the same color) and enough plastic beads to correspond to the number the class is studying. If you're working on the number 7, give 7 beads to each child.

Have each child thread the beads onto her pipe cleaner and then twist the ends of the pipe cleaner together so the beads don't spill all over the floor. Next, have the children separate the beads into sets to create number sentences. If a child has 7 beads on a 7 bracelet, she can separate them into a set of 4 and a set of 3 to show that 4 + 3 = 7. When you tell the children to flip their bracelets over, they'll see 3 + 4 = 7. If they take their 7 beads and separate them into sets of 2 (with 1 left over), they'll show 2 + 2 + 2 + 1 = 7. To show subtraction, students can group all 7 beads together and then slide beads to one side. For example, 7 − 3 begins with 7 beads; then the student pulls 3 beads away.

Estimation Jar

Begin an estimation jar in your class. Each week put a different group of items (e.g., marshmallows, pennies, cotton balls, or cereal) into the jar. Close the lid tightly! All week long, invite students to estimate the number of items in the jar. Estimates can be written on individual cards, in math journals, or on a class chart, or they can be offered as part of a discussion. Let students change their estimates during the week if they feel they can improve on their original predictions.

On Friday, spill the items out onto the floor or table where all can see. Once the children have seen all the items spilled out, give them a chance to change their estimates once more. Discuss. As a class, count by 2s, 5s, or 10s to get the true count of items and let students see how close their estimates were.

For older classes, separate the items into 2, 3, or 4 sets, laying the groundwork for division. "Wow, when we spilled out the 97 cotton balls and placed them in sets of 10, we had 9 sets of 10 and 7 left over. So 97 divided by 10 equals 9 with a remainder or left over of 7. When we put the cotton balls in sets of 5 we had 19 sets with 2 left over. So 97 divided by 5 equals 19 with a remainder or left over of 2." And so on.

Messy or Friendly Numbers

Some numbers are just plain *hard* to compute. For instance, 37 is a tough number to work with. So my class thinks about numbers like 37 as "messy." We stop and discuss what comes to mind that is close to 37 but "easy" or "friendly." "Ah! Yes, 30 and 40 are both clean and neat. Let's try those! If I wanted to add 37 + 17, I could break it apart." (This is the distributive property.)

"I could break 37 into 30 and 7, so I could say 30 + 17 = 47. And then I could say that 47 + 7 [this may take some time] is 40 + 7 + 7, or 40 + 14, which is 54!"

Since doubles (4 + 4, 5 + 5, 6 + 6, 7 + 7, etc.) are usually facts that children memorize earlier, I want them to look for those doubles in the mental math we do together. Students may have a hard time with 47 + 7, but it becomes easier when they convert it to 40 + 7 + 7. Once they look at it that way, most kids will clump the 7 + 7 together because they know it, and then they'll add the resulting 14 to 40. Of course, they need guidance from you to do this.

Think out loud as you solve problems like these: "Whoa! This is messy. There must be some friendlier numbers that I can use to make this easier for me."

Continue: "Now let's try adding 37 + 17 a different way. We know 37 is close to 40, so let's pretend we're starting with 40. You know that 40 + 17 is 57. However, you pretended 37 was 40 by adding 3, so now take the 3 away. In other words, 57 − 3 = 54!"

Standards are listed with the permission of the National Council of Teachers of Mathematics (NCTM). NCTM does not endorse the content or validity of these alignments.

Back to the 100 Chart Game Board

The 100 Chart Game Board is the cheapest, easiest-to-carry math tool you can find. Use the reproducible on page 156 to make a copy for each child. Mount the copies; you may even want to laminate them.

Tell each student, "This is your game board for the year. Take good care of it because it has the answers to countless math problems." Encourage your students to use the game board to solve computation problems.

You'll need extra, unlaminated copies of the game board to help students see the patterns in our number system. Give one of those copies to each student and ask everyone to color with yellow crayon all the numbers in the column that has 1 at the top. Then ask them to use orange to color all the numbers in the column that has 2 at the top. They should use green on the 3s column, and so on. Doing this will help them see what happens to a number when you add 10.

Of course, you can help with the discoveries. "Oh, look at that! In the 1s column every single number has a 1 in the ones place. Gee, in the 2s column every single number has a *2* in the ones place! Now look in the 10s column. Look at the first digit in each number. Those numbers in the tens place increase by 1.

"Wow, this is really cool stuff! We go from 2 to 12 to 22 to 32 to 42 to 52. The numbers in the tens place go 1, 2, 3, 4, 5, but the numbers in the ones place stay at 2! Can you make any other observations from our colored chart?

"Why are all these things true? That's right: because our entire number system is based on patterns!"

While all of this seems so simple to adults, we have to remember that it is not obvious to children. I guarantee that as you do this you will hear oohs, aahs, and wows. They will have made some incredible discoveries that will last a lifetime.

Now, on *another* copy of the game board, have students color in a red, yellow, blue pattern. This time students should change colors with every *square* rather than every column, so the 1 is red, the 2 is yellow, the 3 is blue, the 4 is red, and so on. Guide the children in discovering the patterns within this game board. Note that all the right-to-left diagonals are the same color. Then help students to see that whenever you want to know what's 9 more than a particular number, the answer is in the same color as the original number. Why is this? Because our entire number system is built on patterns!

Coloring a 100 Chart Game Board helps kids discover the patterns in our number system.

You Bet!

Read to the class the book *Betcha!*, by Stuart J. Murphy. In this story, two boys find estimation to be a very efficient and accurate way to predict the total number of everything from cars to jelly beans. The illustrations and text give excellent directions, helping children to see how estimation can be practical and close to accurate.

~~~~~~

EXPECTATION: Develop fluency with basic number combinations for addition and subtraction.

~~~~~~

Review of Facts

When Joseph Renzulli, a recognized leader in the education of gifted students, was speaking to a group of teachers in 2005, he said something that's become one of my favorite quotes: "When the lesson is engaging there is greater achievement." Isn't that a great comment? And you can act on it. Each day set aside time to do a quick review of facts. But don't make it a drill. Instead, follow Joseph Renzulli's cue. Make it a game and you'll make those facts stick.

What kind of game? Try giving your students directions along these lines:

● Roll 2 dice with a partner and see who can add the 2 numbers more quickly. Repeat, keeping score with tally marks.

● Work with a partner. Each of you should flip a card at the same time. See who can call out the total of the 2 cards first. The first child to say the answer keeps both cards. Continue until the teacher calls, "Time's up!" The winner is the player who ends up with the most cards.

● See who can write the best rap or the best poem about adding 9, subtracting 12, or adding a few columns of numbers.

Construction-paper "houses" help students to learn number families. And you can use the houses over and over again if they're laminated.

 # Number Houses

Before class, cut a sheet of construction paper roughly in the shape of a house. Make it at least 9 inches wide and 11 inches from the base to the peak of the roof. In a contrasting color, add 3 windows and a door; each of these should be just slightly larger than a 3 x 3-inch sticky note. Mount the house on a larger sheet of construction paper and laminate the whole thing. Repeat for each student in the class. Then place 4 sticky notes on each house—1 on each window and 1 on the door. From this point on, the directions change according to the ages and abilities of the children.

For younger students: Write 1 number on each of the 3 window stickies. Let's say you use 6, 2, and 8. The child studies the numbers. Then, on the door sticky, he writes the 4 number sentences that use all 3 of the numbers in the windows. In this case, he would write:

$$2 + 6 = 8 \qquad 6 + 2 = 8 \qquad 8 - 2 = 6 \qquad 8 - 6 = 2$$

For older students: Write numbers on only 2 of the window stickies. Let's say you write 76 on one sticky and 54 on another. Each student's job is to fill in the third sticky with a number that represents the total of the 2 (in this case 130) or the difference between them (subtracting one from the other to get, in this case, 22). Then she writes the 4 possible number sentences on the door sticky. Depending on skill levels, at some point you may want to make this more challenging by using 3-digit numbers.

Once students have worked with the houses and understand the basic idea, they may create all 3 windows, *and* the door, on their own.

EXPECTATION: Use a variety of methods and tools to compute, including objects, mental computation, estimation, paper and pencil, and calculators.

Animal Party

In *Bear Snores On*, a darling story by Karma Wilson, a bear sleeps in his den while a variety of other animals creep into his cave for protection from a storm. As more and more animals arrive, a small party develops. In the end the bear wakes up, frightening the other animals. However, all he wanted was to be part of the party himself.

The text in this book is as delightful as the illustrations, so first read the book to your students for pure enjoyment. After that initial reading ask, "How many animals are in this story?" Gather estimates from the class, accepting all possible answers.

Standards are listed with the permission of the National Council of Teachers of Mathematics (NCTM). NCTM does not endorse the content or validity of these alignments.

53

Then proceed along these lines:

- Say, "Let's try different ways to count the animals in the story. I'm going to read this story again." Point to a third of the class and say, "Everyone in this group will hold up one finger each time a new character enters the bear's cave."

- Point to the middle third of the class and say, "Everyone in this group will make one tally mark for each animal." Hand out writing supplies to each student in that group.

- Point to the last third of your class and say, "Everyone in this group will make a tower by adding one Unifix cube for each new animal." Distribute cubes to each child in this third.

- Read the story again, checking that each group is counting correctly.

- After you finish reading the story, compare the answers from the 3 groups. Be *delighted* that each method resulted in the same answer.

This is a wonderful way to demonstrate that often there are many different ways to solve the same problem.

How Many Critters?

My Little Sister Ate One Hare, an outrageously funny book by Bill Grossman, has the key elements of great literature for children: the author mentions underwear and throwing up! The kids will be entranced as you read the story. Soon they'll be finishing the lines on each page as you read, showing that they understand the pattern of the story.

For this activity, follow steps something like these:

- Read the story to the class, pausing for certain laughter. (Plan on a few minutes of hysterical laughter when you get to the underwear page.)

- After that first reading ask, "How many critters did the little sister eat?" Be certain to clarify that she ate 1 thing, then 2 things, then 3 things, and so on.

- Be clear in explaining, "It doesn't matter *what* she ate. I'm asking you *how many* she ate."

- Give children the freedom to solve this problem using manipulatives, pictures, tally marks, or numbers. Praise independent thinking! Say, "I'm going to give you ____ minutes to solve this problem. We'll have a math meeting to compare our answers and our strategies in ___ minutes."

- Add, "If you happen to finish early, remember that good mathematicians *always* check their answers. I'm looking for extra-smart mathematicians who will try checking their work using a different strategy."

- Turn them loose to problem-solve. This is your chance to kid-watch, noting who uses cubes, who counts on fingers, and who prefers pictures, tallies, or numbers. Their strategies are windows to their understanding; you can use this information as you plan later lessons.

- When time is up, call children to the math meeting.

- Ask, "How many critters did she eat?"

- Write down the answers your students give. Don't give any facial clues as to whether the answers are correct or not. Don't be concerned if there are some wrong answers.

- Ask, "Who would like to justify his or her answer?" Allow students to explain their problem-solving processes.

- If no one chose a T-chart say, "I'd like to show you a way to solve this problem with a T-chart. Watch how I use the T-chart."

- Demonstrate making the T-chart. Write "Number She Ate" on the left side of a sheet of chart paper and "Total She Ate" on the right side. Underline those labels. Draw a vertical line separating the 2 labels and continuing down the rest of the paper.

- Go line by line, discussing each line with the class as you add it: "Okay, next she ate 4 disgusting things, so I'll put 4 in the left-hand column. She already ate 6 critters, so in the right-hand column I'll put 6 + 4. We know that 6 + 4 = 10 critters total, so I'll put that answer,10, in the right-hand column, too."

- Emphasize that this is *one* way to solve the problem.

- At the bottom of your chart, write something along these lines: "We read *My Little Sister Ate One Hare*. We figured out how many critters she ate. Some of us used Unifix cubes, some used blocks, some used ducks, some used tallies, some used numbers, and some used T-charts. We used different methods, but we all got the same answer."

Number She Ate	Total She Ate
1	1
2	1+2=3
3	3+3=6
4	6+4=10
5	10+5=15
6	15+6=21
7	21+7=28
8	28+8=36
9	36+9=45

This problem gives you a way to introduce T-charts.

How Old Is This Class?

Begin by asking, "If we add up the ages of all the children in this class, how old would this class be?" Let children suggest different ways to solve the problem, and then continue along these lines:

- Ask, "How can we discover how old the youngest person in this class is? How can we discover how old the oldest person in this class is?"

- Do a quick survey of the class, asking questions like, "How many of you are 6 years old?"

- Post the results of your survey where everyone in the class can see them.

- Turn this problem over to your little problem-solvers, giving them a time frame in which to work. Say something like, "Okay, problem-solvers, you have _____ minutes to solve this.

Before students can figure out how old their class is, help them come up with some data they can work with.

You may solve this problem in the way that works best for you. After ___ minutes we'll have a math meeting to discuss our results. Remember that good mathematicians always check their answers!"

- As they're solving the problem, take the opportunity to kid-watch, observing the strategies that children are using.

- After the designated number of minutes announce, "All right, mathematicians, meet me at the easel and let's discuss how you solved this problem."

- If you're lucky, your students will have come up with a variety of ways to solve the problem. Celebrate this diversity by asking students to explain their strategies.

- With the children dictating and you as their scribe, write a summary of the activity on chart paper. Use correct mathematical language. You might write something like this: "We added all of our ages and learned that this class is _____ years old! We used cubes, keys, sticks, tallies, and numbers to figure this out."

- Ask your students to record in their math journals what they did and what they learned.

  Note: Depending on the makeup of your class, this could cause hurt feelings on the part of a child who is older or younger than classmates. Please consider this before planning the activity.

Watch to see what strategy a child uses to solve this problem.

● Candle Count

This is another way to reinforce the idea that many different approaches can all lead to the same answer.

- Begin by asking, "How many birthday candles have you blown out in your lifetime? That's kind of a tough problem to solve. So before we call out numbers, let's think of some efficient ways to solve this problem."

- Say, "Let's think about this. When you were 1, you blew out 1 candle. When you were 2, you blew out 2 candles. When you were 3, you blew out _____." Let children fill in the answer of 3 candles. "Does anyone see a pattern here?"

- Keep the discussion going.

- "I want you to figure out how many candles you have blown out in your lifetime. You can use any method you'd like. Once you've solved this problem, prove your answer. We'll have a math meeting in ___ minutes and we can discuss the answer then."

- Send the candle-blowers off to solve the problem, and then gather them into a group at the designated time.

- Ask, "Will everyone in this classroom have the same answer? Why? Or why not?"

- Allow discussion of the different numbers of candles that have been blown out and why the numbers are different.

- Take time to celebrate the fact that there are many ways to solve this problem. Let your students know that you are proud of them for thinking of different ways to figure out the candle count.

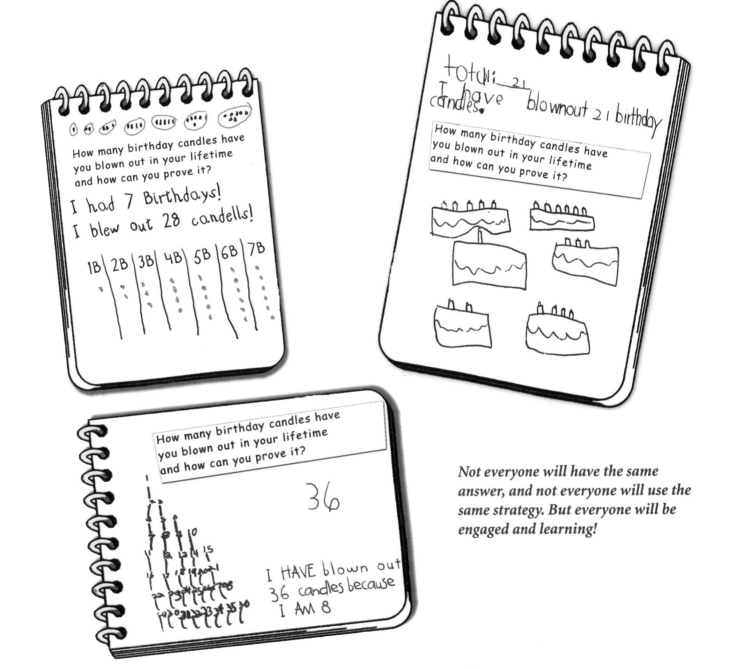

Not everyone will have the same answer, and not everyone will use the same strategy. But everyone will be engaged and learning!

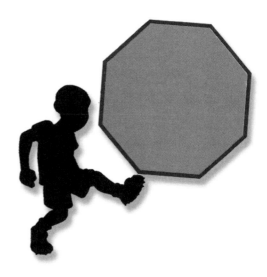

ALGEBRA

For many of us, the word "algebra" conjures up recollections of terror, memorization, and bewilderment. As students we sometimes wondered, "When on earth am I going to use this?" Actually, we use algebra every day. You are probably using it without even knowing it. When you watch the trends in mortgage rates, select the perfect patterned wallpaper, make generalizations, and read charts and tables, you're using algebra. Have you ever gone to the store with x amount of money and wanted to buy something that cost almost x? And then you needed to know how much money you'd have left because if it was more than y, you could also buy that extra-special something more? You solved an algebra problem to figure that out!

For years algebra was considered the work of middle- and high-school teachers, but we now know that our kindergarteners can handle algebraic lessons, too. In the late 1990s, after observing children in many classrooms from kindergarten through grade 6, researchers Joan Ferrini-Mundy, Glenda Lappan, and Elizabeth Phillips concluded that children were capable of mathematical insights and inventions that exceeded conventional expectations of the time (Ferrini-Mundy 1997). Primary-school students were noticing patterns and creating tables to solve problems. Our students, we learned, can handle this! What's even better is that by introducing algebra in the primary grades, we are making children's study of mathematics in middle school and high school richer.

Traditional classes focused on the right answer. Today the bar has been raised. The National Research Council (NRC) has stated: "Traditionally instructed students who are proficient with numbers need to shift from thinking about 'finding the answer' to thinking about the 'numerical relationships' underlying the calculations they perform and the nature of the methods they use" (Kilpatrick 2001). In other words, we don't want kids to become automatic robots mindlessly adding and subtracting numbers. Rather, we want to encourage thinkers who know what they are doing and why they are doing it and who see relationships in their answers.

Talk to your students about the generalizations they can see in problems. "That is so cool how every time we add a 9 to a number it is the same as adding 10 and taking away 1." Yes, that's number sense, but because you are making generalizations, it is also algebra. Or you might say, "Isn't that interesting that every shape with 3 angles has 3 sides, and every shape with 4 angles has 4 sides?" Yes, this is geometry, but again it is also algebra; once again you are making a generalization.

Algebra is the study of patterns. Red fish, blue fish, red fish, blue fish, and 123, 123, 123 are patterns. Children at the art center stamping patterns with sliced fruit are thinking algebraically. When you and your students sing, clap, and recite rhymes, you are also doing algebra. "Alligator Pie," "Row, Row, Row Your Boat," and "Do Your Ears Hang Low?" are chock-full of patterns. So sing your heart out. It's more than music; it's also algebra!

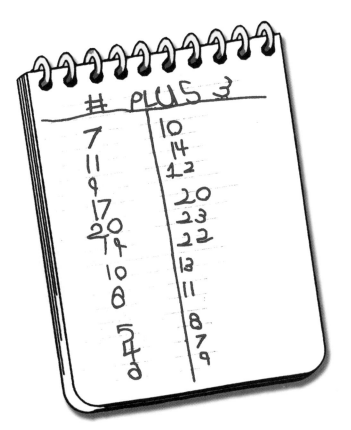

T-charts help students to organize their thinking and see patterns.

STANDARD Understand patterns, relations, and functions.

EXPECTATION: Sort, classify, and order objects by size, number, and other properties.

⬡ Where Does This Belong?

Ask each child to give you one shoe. (All will laugh. Some will dramatically cover their noses and gag. Kids are real actors!) Pile the shoes by your side. Seat single-shoed children in a circle around you. Start sorting shoes into 2 separate sets, Velcro and no Velcro—but do not tell the children *how* you are sorting the shoes (a.k.a. the sorting rule).

Once you have placed 3 or 4 shoes in each set, ask the class if they can tell you how you sorted the shoes. If they are not able to identify the rule, continue sorting until a child can explain the rule. Once the rule has been determined, discuss the attributes of the sets. Ask, "Which set has more?" "Which has fewer?" "What is the difference in the number of shoes in these sets?" "How many Velcro shoes are there?" "How many no-Velcro shoes are there?"

Use the same shoes to sort again. This time sort by "white soles" and "not white soles." Again, do not tell the class your rule; let them figure it out. Continue with the same sort of questions you used for the Velcro sort. Ask, "Can any of you smart mathematicians come up with another sorting rule?"

Another type of sorting deals with numbers: "All buttons with 2 holes go over here; all buttons with more than 2 holes go over here." Or, "If you have 4 members in your family, stand here; if you have fewer than or more than 4 members in your family, stand over here." Or, "All kids with zero pockets stand in this line, children with 1 pocket in this line, and those with 2 pockets stand here. If you have more than 2 pockets, please stand here."

PATTERN-SEEKERS BECOME PROBLEM-SOLVERS

Patterns are everywhere! They appear in our clothes, the shapes on our wallpaper, the ceilings of our classrooms, the sidewalks we walk on, and even in our number system. In fact, our number system is *based on* patterns. We want our students to become pattern-seekers so that they will clearly see the patterns in numbers. Understanding the patterns in the base-ten system enables children to become strong mathematicians and to gain confidence in math because they realize that there is a pattern behind the answers we get when we add, subtract, multiply, divide, and problem-solve.

Standards are listed with the permission of the National Council of Teachers of Mathematics (NCTM). NCTM does not endorse the content or validity of these alignments.

61

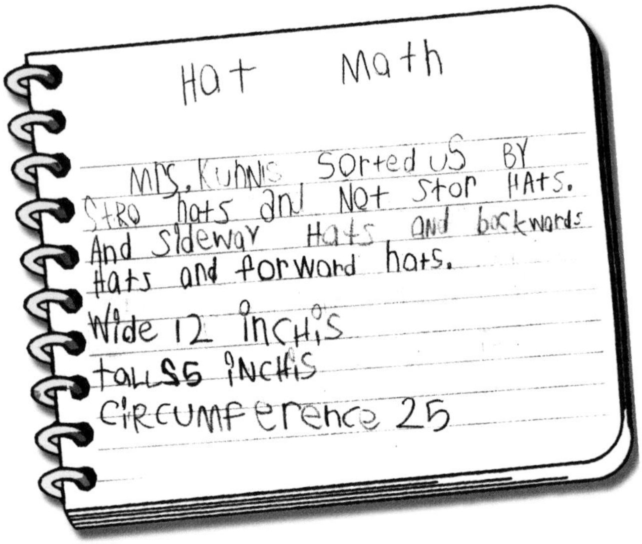
Hat Math
Mrs. Kuhns sorted us by
Stro hats and Not Stor HAts.
And Sideway Hats and backwords
hats and forword hats.
Wide 12 inchis
talls5 inchis
circumference 25

Hats off! After students sort their hats, they can measure them too.

Hat Day

Ask kids to come to school in their favorite hats. Begin the day by reading *Caps for Sale,* by Esphyr Slobodkina, to the class. After that, you could spend all day sorting kids by their hats! Think of the possibilities: straw vs. cloth; sports teams; colors; turned backward or not; pictures, words, or both; using printing, cursive, or neither; with brims and without brims.

Flip Your Lids!

Lids are wonderful for sorting! They have countless sorting possibilities. Ask families to send in lids and/or caps from shampoo, milk, jelly, margarine, salad dressing, peanut butter, orange juice, and anything else they have on hand. Lids and caps can be sorted by letters and no letters, pictures and no pictures, small and large, metal and plastic, and so on. The kids will think of ways you haven't even considered.

Never Be Out of Sorts

Here are some other great materials for sorting:

- **Buttons**
- **Cereal**
- **Mittens**
- **Toothpaste boxes**
- **Cereal boxes**
- **Beads**
- **Seashells**
- **Plastic cups**
- **Beans**
- **Blocks**
- **Key cards from hotels**
- **Plastic silverware (by colors, by type)**
- **Stuffed animals (by kind, by clothes, by colors, by eyes open/shut, by ribbons/no ribbons, by size, by sitting/not sitting)**
- **Paper plates (The kind from party stores come in all colors, sizes, and designs. Ask families to send in any plates that are left over from parties.)**

Kid Sorts

The best manipulatives are the kids themselves. Kid Sorts are a great way to fill a few minutes at the end of the day with fun and content. Your students will come up with great ideas for sorting themselves. To get them started, try any of these sorting rules:

- T-shirts: Sort by collar and no collar, words and no words, pictures and no pictures, pockets and no pockets.
- Pants: Sort by jeans and not jeans, short and not short, pockets and no pockets.
- Hair: Sort by color, curly and not curly, ponytail and no ponytail, hair scrunchies and no scrunchies.

Sort It Out!

After the children have watched and helped you sort a variety of items, it's time to turn things over to them. Your job is to pass out the materials to be sorted. Their job is to decide the sorting rule and then sort. All too often, well-meaning teachers pass out the objects and say, "Okay, now sort these fabric swatches by red designs and no red designs." That's so wrong!

When you determine the sorting rule, you take part of the math out of your lesson. Instead, give your students the opportunity to decide the sorting rule, let them sort, and then let them tell their tablemates or the class the rule. Even better, let them write their rule in their math journals as a perfect closure to the lesson. On another day, do it again. Any lesson like this needs to be repeated many times to be sure students have really got it!

A SCIENCE EXPERIMENT

When we leave for lunch or a special class, I always say something like, "I'm conducting a scientific experiment. I want to see if children with short sleeves are quieter than children with long sleeves. I'm watching this line all the way to music." I change the attributes each day and my classes always get into this classification contest. They even come up with the sorting rule. Funny, it is often a tie and the winners can be determined only by the walk *back* to class.

Storybook Introductions

As a springboard to sorting, read *The Button Box* and *A String of Beads* to the class. Written by Margarette S. Reid, both include beautiful illustrations that show sets of beads and buttons sorted by different attributes. Make sure the children are seated close to you to appreciate the illustrations. These books will generate great discussions about grandmas and varieties of buttons and beads.

Introducing Venn Diagrams

Once they understand sorting rules, first and second graders can start using Venn diagrams to sort for the intersecting set. Call the children close to you; ideally, have everyone seated on the floor. Bring with you a collection of jar lids. Place red lids to your right. Say, "These lids are all red." Place metal lids to your left. Say, "These lids are all metal." Then look puzzled and say, "Oh, dear! Where will I put these red metal lids?" Pause. "I know. The lids that are red *and* metal belong in the center. They are the intersection of the two sets. Mathematicians say that items that fit both set rules belong in the *intersecting set.*"

Point to the red set. "This is the red set." Point to the metal set. "This is the metal set." Point to the set in the middle. "These lids are red *and* metal. I call this set the intersecting set."

To make this very clear, draw circles around each set, making the two circles overlap to show the intersection of the two sets. I usually do this by drawing chalk circles on the carpet. (It vacuums right up.) If you can't draw the circles, use Venn sorting

Standards are listed with the permission of the National Council of Teachers of Mathematics (NCTM). NCTM does not endorse the content or validity of these alignments.

63

hoops, available from teacher stores and education catalogs, or lay two hula hoops on the floor to create a Venn diagram. Continue sorting with two attributes and an intersecting set.

Universal Set

Once your students understand sorting rules and intersecting sets, introduce the concept of the universal set. Using hula hoops, chalk lines, or plastic Venn rings, again place those red lids in one set and metal lids in a second set. Again place the red metal lids in the intersecting set. Say, "These lids are red. This is the red-lid set. These lids are metal. This is the metal-lid set." Point to the intersecting set and add, "These lids are red *and* metal. They are the intersecting set." Now point to some plastic lids cast off to the side. Say, "These lids are not red and not metal. So they do not belong in either set. Mathematicians call these gold plastic lids part of the *universal set* because they don't fit in anywhere!"

Use jar lids to explain the concept of the universal set.

EXPECTATION: Recognize, describe, and extend patterns such as sequences of sounds and shapes or simple numeric patterns and translate from one representation to another.

Create Manipulatives Patterns

Use manipulatives like keys and pennies to assemble a pattern in front of the children. Say, "See, this pattern is key, penny, key, penny, key, penny." Ask children to say the pattern with you. Then add, "I can also call this pattern K, P, K, P, K, P. I'm using K for key and P for penny. Or I could say this is an A, B, A, B, A, B pattern. Or I could call it silver, copper, silver, copper, silver, copper. Can anyone think of another name for this pattern?" Allow children to give your pattern a new name. It is very important for children to understand that the different words and letters all represent the same pattern.

Repeat the A, B, A, B pattern using different materials. Then give students different manipulatives and say, "I want you to use these manipulatives to create patterns that are A, B, A, B, A, B." Once they've created the patterns, ask each child to name his pattern.

Sounds Good

Create patterns with sounds. Let's say a child creates a pattern of plastic creatures. Maybe her pattern is dinosaur, bug, frog, dinosaur, bug, frog. You might say, "This is an A, B, C pattern or a dino, bug, frog pattern. It could also be a clap, slap, snap pattern. Let's try that. We'll clap when I point to the dino, slap our thighs when I point to the bug, and snap our fingers when I point to the frog." Do this several times.

Say, "We can also give each critter a sound. This can be a roar, buzz, ribbit, roar, buzz, ribbit pattern."

Once the class understands the routine, turn this over to your students: "Okay, who would like to decide the theme and the pattern?"

MUSICAL PATTERNS

Sound patterns are especially appealing to auditory learners. Besides, finding patterns in sounds helps to demonstrate that patterns are everywhere. Certainly you and your students can find plenty of patterns in music, too. A rather famous musician, a fellow named Mozart, composed many pieces that had patterns within patterns. My students enjoy it when I play classical music during work time or at the end of the day. I tell them, "Listen for the patterns Mr. Mozart put in his music." And they do!

Sound Patterns

Gather students in a circle with everyone facing the center. Say, "We are going to do a farm-noise pattern. I'll say 'quack.' The next person will say 'hee-haw,' and the next person will say 'baaaa.' This is a 1, 2, 3 pattern or an A, B, C pattern. Pay attention so that when the pattern comes around to you, you will know what to say." Paying attention is a tough task for some, but they'll get it—or their friends will help them out!

You can use this activity with jungle noises, robot noises, spooky noises, or whatever you or the students want. Ask, "Can you think of other sounds that go with the pattern?"

Body Patterns

Again bring students together in a circle, with everyone standing and facing the center of the circle. Say, "Okay, now we're going to do an up, up, down, up, up, down pattern. The up people are going to stand on their toes. The down people are going to bend their knees and put their hands on their knees. I'll start. See how I'm standing on my toes? Now keep the pattern going around the circle."

Once they get the idea, continue Body Patterns using kid-inspired positions.

Once your kids understand body patterns, they'll come up with plenty of ideas of their own!

Describe Pattern Units

Explain, "Mathematicians have a special name for what's inside your pattern. If your pattern is cat, dog, cat, dog, cat, dog, then you have cat and dog in one unit of the pattern. This pattern has 2 members in the unit—1 cat and 1 dog. That makes 2!"

Children will need lots of practice with this concept, so point out the units each time you create patterns together. Maybe a child has pattern blocks lined up something like this:

You might say, "I see your pattern shows 4 units and you have 3 members to a unit." Or you might ask, "How many units are in your pattern? How many members are in each unit?" It will take lots of discussion and practice, but your students will learn to start using the new lingo.

Make sure everyone understands that the unit is all the elements (pieces, shapes, blocks, numbers, sounds, or whatever) of the pattern that appear before the repetition begins. So if the pattern is A, B, C, A, B, C, it has 3 members to the unit: A, B, and C. And it has 2 units because A B C appears twice. If the pattern is blue square, blue square, yellow circle, blue square, blue square, yellow circle, it has 3 members in a unit. The unit is made up of 2 blue squares and 1 yellow circle. And it has 2 units.

Point out other patterns and ask the class to tell how many units are in each pattern and how many members are in each unit. "Let's look at this pattern. It's all frogs: green, green, orange, blue, green, green, orange, blue. How many members are in a unit? That's right, there are 4: green frog, green frog, orange frog, blue frog. How many units do you see? That's right. You see 2 units!"

Have children create patterns. Then ask them questions comparing their patterns and their friends' patterns. You want them to tell you how the patterns are alike and how they're different. Make sure the questions necessitate answering in sentences, never just "yes," "no," or "uh huh."

A Quick Tip

Cheerios patterns are challenging since all Cheerios look the same. How about grouping the cereal in sets of 2, 3, 2, 3, and so on? Or lay the cereal on a table in a horizontal line, then move every other piece up above the line. That gives you a pattern of Cheerio down, Cheerio up, Cheerio down, Cheerio up.

Create a Pattern

Ask, "Can you create an A, B, A, B, A, B pattern?" "Can you create a pattern using these pattern blocks?" "Can you create a pattern using Froot Loops? Trix? Cheerios?"

Extend Patterns

Encourage children to look closely at patterns. Ask, "How can you extend this pattern?" "How many units do you see to this pattern?" "Can you add 2 more units to this pattern?"

PATTERN TYPES

There are 2 types of patterns: repeating and growing. Patterns that repeat are called (surprise!) repeating patterns. We see repeating patterns all over the place. Repeating patterns have units, as in the example at right.

Growing patterns are patterns that (surprise again!) grow. Look at 1, 2, 3, 4, 5. That's a pattern that grows by 1. Or look at 2, 4, 6, 8, 10, which is a growing pattern that grows by 2. Another growing pattern is 1, 3, 6, 10, 15, 21. It grows by a pattern of $+ 2 + 3 + 4 + 5 + 6$. Growing patterns do not have units.

EXPECTATION: Analyze how both repeating and growing patterns are generated.

Chart It!

Children need to understand that there are a variety of ways to solve a problem and that creating charts is one very good way to organize and solve problems (NCTM 2000). When you are problem-solving with your class, show them a variety of methods to find the answer. Name the methods you are using.

Now this may come as a real surprise, but children are not naturally organized. Clear, easy-to-follow charts help children keep information from getting jumbled. As I watch my early first graders copy everything I do, I know that soon they'll launch into a T-chart or a table with no prompting from me. By second grade they'll really take the ball and run with it.

For example, let's say you want to determine the number of wheels on a bunch of tricycles. The pattern is counting by 3s. A simple table is easy to follow and a good way to solve the problem.

WHEELS ON TRICYCLES	
Number of Tricycles	Number of Wheels
1	3
2	6
3	9
4	12
5	15
6	18
7	21

TOES ON KIDS	
Number of Kids	Number of Toes
1	10
2	20
3	30
4	40
5	50
6	60
7	70

What if you want to count the number of toes on kids? In that case, the pattern is counting by 10s. A T-chart works well.

And if you want to count the number of paws on puppies? In that case, the pattern is counting by 4s. A table like the one below helps to clarify the information.

These are different types of charts or tables. It is important that the children see a variety of charts and understand that more than one way of charting a problem is possible. It is also important for children to realize that an efficient and organized manner of counting and problem-solving is the most reliable. After all, how can you go back and check your work if it is scribbled with no rhyme or reason?

PUPPY PAWS							
Number of puppies	1	2	3	4	5	6	7
Number of paws	4	8	12	16	20	24	28

Double Delightful

Read *Two of Everything*, by Lily Toy Hong. This Chinese folktale is a perfect introduction to doubling. After reading the book to the class, ask, "What if I had a magic pot that would double everything I put into it? If I started with 1 gold coin, how many tosses in the pot would it take to have 8 gold pieces? Can we solve this in a T-chart?"

Then challenge your little Einsteins with, "How many tosses into the pot will it take to get more than 30 gold pieces?"

Then ask, "How far can you take doubling on our chart?"

Number of Tosses	Number of Gold Pieces
1	2
2	4
3	8

A T-chart introduces the concept of doubling.

Another way for students to practice doubling is by "Doubling the Dinner." This is real-world math!

Doubble the dinner
36 friends   72 friends

9 Bags Carrots	18 Bags Carrots
36 Lemons	72 Lemons
3 SD	6 SD
6 lb Cucumber	12 lb Cucumber
9 Letuce	18 Lettuce
18 Tamatos	36 Tamatos
2 Onions	4 Onions

Grow the Playground Pattern

Ask children to solve this problem:

> Pat is painting the playground with a special pattern, but she lost the directions. Look at the pattern below and write a letter to Pat telling her how to continue the painting and how you figured it out.
>
> O OO OOO OOOO OOOOO _____ _____

The answer, of course, is OOOOOO OOOOOOO, so the letter would read something like this: "Dear Pat, You need to add 1 circle each time you start a new set. You are painting a growing pattern."

Grow the Houses Pattern

Ask students to solve this problem:

> George has finished painting the houses on Clark Drive. He needs to replace the address numbers but he knows the addresses on only 2 of the houses. Write George a letter letting him know the missing numbers and how you figured it out.
>
> 161 _____ 171 _____ _____ _____ _____

The student's answer should read something like this: "Dear George, You should count by 5s starting at 161, so the houses will be numbered 161, 166, 171, 176, 181, 186, 191. You are painting in a growing pattern."

Standards are listed with the permission of the National Council of Teachers of Mathematics (NCTM). NCTM does not endorse the content or validity of these alignments.

69

Bring Back the Game Board

The very same 100 Chart Game Board discussed in the Number & Operations chapter is useful in this strand as well. After all, our number system is based on patterns. Have each child highlight all even numbers on a copy of the game board. Guide students to see that this also points out the odd numbers. Then have them highlight all numbers that have a zero in the ones place, and you can help students focus on skip counting by 10 and on products of 10.

STANDARD

Represent and analyze mathematical situations and structures using algebraic symbols.

EXPECTATION: Illustrate general principles and properties of operations, such as commutativity, using specific numbers.

Today's Date

It is important for students to understand that since $5 + 6 = 11$, then $6 + 5 = 11$. This demonstrates commutativity. Knowing and truly understanding this will help them also realize that $11 - 6 = 5$ and $11 - 5 = 6$.

Remember the Dose of Daily Digits strategy (pages 35–37)? Let's go back to the calendar now. Using the numbers in the date will give students daily practice and help them internalize these concepts. If it is September 5, or 9/5, then reinforce this concept by showing $9 + 5 = 14$, $5 + 9 = 14$, $14 - 9 = 5$, and $14 - 5 = 9$. (For younger students, you may need to use simpler numbers instead of the exact date.)

During your calendar time or as a wrap-up to a math lesson, point out the numbers in the date and quickly review the two related addition sentences for those numbers. Write the number sentences on a chart or the board. Say, "Okay, today is September 8th, or 9/8, so let's work with 9 and 8. Who can finish this number sentence?" Write on the board or chart: "$9 + 8 = $."

Say, "$9 + 8 = $ what? Yes, it's 17. Now watch this." Write "$8 + 9 = $." Ask, "What is 8 plus 9? That's correct: 17. Both $9 + 8$ and $8 + 9$ equal 17. When we add, it doesn't matter what order the numbers are in; the answer will always be the same."

If you do this as a fast review on a daily basis, it will prove well worth the few minutes you spend on it.

Variation: As children's skills and confidence grow, move away from the calendar and select numbers that are more of a challenge—maybe the number of girls in the class and the number of boys.

Algebra Triangles: Seashells on Lakes

For children who understand addition, algebra triangles are a great way to problem-solve and to see the relationships in addition number sentences. For each child, you'll need 3 small work mats (created from construction paper, using the reproducible patterns on page 159), 3 craft sticks labeled with markers, and enough small manipulatives to represent the facts you want the child to work on. Before class, you'll need to make the work mats and label the craft sticks. Using the photos on this page as models, write a single-digit number on one side of each craft stick in green, and then turn the stick over and write a double-digit number on the back in blue.

To introduce the algebra triangles, sit on the floor and ask the class to stand around you as you demonstrate. You want them to hover above you as you explain so they can get a bird's-eye view of how this is done.

Here's how the strategy works: For each student, you place the 3 lakes in a triangle formation and label them as A, B, and C. The child takes a craft stick that's labeled with an A on one end and a B on the other. He places the A end of the stick on Lake A and the B end on Lake B, keeping the green-numbered side of the stick facing up. He positions the other 2 sticks similarly, matching the letter at each end of the stick with the letter on the corresponding lake. Then he looks at the number in the middle of each stick.

Let's say the number on the stick that connects A and B is 6, the number on the stick that connects B and C is 9, and the number on the stick that connects A and C is 7. The child needs to figure out how many shells go in each lake so that when he adds the contents of Lake A and Lake B, he gets a total of 6—matching the number on the craft stick. And of course he needs to make the total for B and C equal 9 and the total for A and C equal 7.

After manipulating the pieces, the child places the shells appropriately and determines that A = 2, B = 4, and C = 5.

But he's not through yet! On the back of each stick is that *two*-digit number written in blue. Once the child figures out the parts for the one-digit totals (all

Algebra triangles help students see the relationships in addition number sentences.

If A + B = 6, A + C = 7, and B + C = 9, then A = 2, B = 4, and C = 5.

in green), he can flip the sticks over and try the more challenging blue problems. Maybe A + B = 17, A + C = 15, and B + C = 16. The child needs to determine that A = 8, B = 9, and C = 7.

Once you've modeled the process, turn students loose in small groups to problem-solve. Don't be too quick to swoop in and help them. They can do this, and they need time to figure it out. You can provide the old "I-have-confidence-in-you" mantra, but let the kids figure things out for themselves.

Algebra Triangles: Caterpillars on Leaves

Before class, copy the reproducible patterns on page 160, cut out the pieces, trace the pieces onto 2 shades of green construction paper, and cut out the "leaves." You need to make 3 leaves for each triangle. For manipulatives, cut pipe cleaners into pieces to represent caterpillars. Print numbers and letters on craft sticks as above. Label one leaf A, one B, and one C. In class, distribute the leaves, caterpillars, and craft sticks and proceed as above.

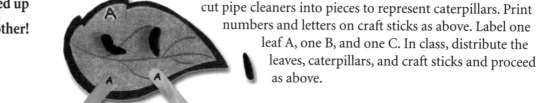

Cut-up pipe cleaners are the "caterpillars" for these triangles.

Algebra Triangles: Ponds

Before class, for each triangle you'll need to copy the reproducible lily-pad patterns on page 161, cut out the pieces, trace them onto green construction paper, and cut out the construction-paper pieces. Copy the pond pattern, cut it out, and trace it onto shiny wrapping paper to represent water. Print numbers and letters on the craft sticks as above. Label one pond A, one B, and one C. In class, distribute the ponds, the sticks, and plastic frogs or flies for the children to use as manipulatives. Proceed as above.

What do you use as manipulatives with ponds? Flies, of course!

STANDARD

Use mathematical models to represent and understand quantitative relationships.

EXPECTATION: Model situations that involve the addition and subtraction of whole numbers, using objects, pictures, and symbols.

Animal Problems

For this one, you'll need something to represent animals. You can use plastic animals. Depending on the word problem, you may be able to use a large block to represent each animal and place smaller blocks close to the large block for ears and paws. Or you can have each child make a simple drawing—a circle for the animal's face, another circle for the body, and (as you proceed with your questions) appropriate numbers of stick legs, triangle ears, and stick tails. Whatever you use, hand out the manipulatives or drawing paper, make sure the "animals" are ready, and then start asking questions.

Let's say you ask a cat question: "Today we have 2 cats. How many ears do they have?" Students can count the ears on their manipulatives, they can add ears to their simple drawings, or they can make tally marks inside those drawings. Ask for answers, and *ask how students arrived at their answers.* Be sure to celebrate the different approaches your students used to solve the problem.

Then ask, "And how many paws do those 2 cats have?" Repeat the process. This points out the relationships among numbers. Later students will think of 2 cats and understand that 2 + 2 = 4 ears and 4 + 4 = 8 paws.

As you read each word problem to your class, write the corresponding number sentence on the board. This will help your students see the connection between word problem and algorithm. Maybe the word problem says, "There are 8 cows in the pasture and 3 more cows join them. How many cows are there in all?" As you read the problem, write "8 + 3 = ." Writing the algorithm while you say the problem helps the child understand the connection. Once the students have given their answers, fill in the 11.

You'll come up with lots of other animal questions to encourage problem-solving in a variety of ways. Here are some examples to get you started:

● There are 4 turkeys and 3 kittens in the barn. How many feet are there in all?

- There are ducks and turtles in the pond. There are 20 feet in all. How many feet belong to ducks? How many feet belong to turtles? Is there only one answer?

- There are 2 ducks, 3 chickens, 1 rooster, 4 horses, and a farmer in the barn. How many feet? How many heads? How many tails? How many ears? How many birds?

A QUICK TIP

I tell my students that the "<" and ">" symbols are like mouths that are hungry for yummies. Because they're hungry, the mouths always open in the direction of the larger number so they can eat more. The point always faces toward the smaller number.

 ## School Days

When you discuss the fact that you've been in school 80 days and there are still 100 days left, write "80 < 100" and "100 > 80." Use the language *as you write* the expression. Say, "80 is less than 100" as you write "80 < 100." Say, "100 is greater than 80" as you write "100 > 80." Write the symbols on the board or a chart whenever you compare numbers. This repetition will help your students become fluent with the symbols "<" (less than) and ">" (greater than).

STANDARD   Analyze change in various contexts.

EXPECTATION: Describe qualitative change, such as a student's growing taller.

 ## Use Comparative Language

This is a very easily accomplished algebra expectation. Yes, I did write "easily" and "algebra" in the same sentence! Meeting this expectation relies on your seizing the moment and engaging your children in conversations that compare change over time.

Model comparative language for your students. Say things like, "There was more water in our bowl yesterday than there is today." Or note that "the bean plant is taller today than it was last week." Encourage children to use qualitative language ("more," "less," "taller," "shorter," "heavier," "lighter") to compare change over time. Algebra involves noting trends and making generalizations. So when you model that comparative language, you are modeling algebra!

EXPECTATION: Describe quantitative change, such as a student's growing two inches in one year.

 ## Use Quantitative Language

We want our students to know that we can take our *qualitative* observations and describe them in mathematical terms. That's the *quantitative* part. Model for your students how they can use language to convey specific measured changes: "Look! The plants are 1 inch taller today than they were yesterday," or "You are one-half inch taller this month than you were in September." Again, encourage them to follow your lead. When you do this, comparing measured change over time, you are modeling quantitative language. Once again, this is algebra!

GEOMETRY

Your students come to school with lots and lots of geometric and spatial knowledge. When they were babies they had shape toys. When they were toddlers, they played with bowls from the lowest kitchen cabinets. For years they've stacked cans in grocery carts. And you know that if you give any child a bunch of blocks, that child will start building right away without any prompting from you. This topic is one that kids love, so you'll have their attention right away. They'll also love creating the projects that they'll design in this chapter, not to mention the beauty of the finished projects. And so will you!

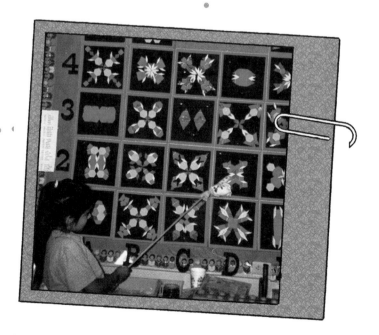

If we are to meet NCTM standards, the emphasis in primary classrooms must be not just on naming the shapes, but also on recognizing the attributes of those shapes. Rather than saying, "This is a triangle," students should learn that "this shape has 3 sides and 3 angles. All the sides seem to be the same length." Rather than just saying, "This is a trapezoid," they should understand that "this shape has 4 sides. Two of those sides look like they are the same length and the top and bottom sides are not the same length." Terms like *tetrahedron* or *truncated pyramid* are wonderful to say and kids love big words, but if the students do not know what makes the shape a truncated pyramid or a tetrahedron they are missing valuable mathematics. This language does not come quickly.

We also want our students to understand how these attributes compare and contrast to one another. "This square has sides that are the same length as the sides of the triangle. The rectangle and square have right angles. This triangle has one right angle and this triangle has no right angles." This can take some time, but with your modeling and their practice, your little charges will become expert at it. And when children can correctly apply those terms, they're ready for some more serious geometry. So let's give it to them!

Borrow building blocks in a variety of shapes and sizes, gather all your pattern blocks, send home letters requesting quilts, check out the many beautiful quilt-themed picture books, encourage families to send in magazines showing homes and gardens, make friends with the produce manager at your supermarket, and put out notice that you welcome any postcards of beautiful buildings. Your classroom will be ready, and you and your students will love discovering the beauty of geometry just about everywhere.

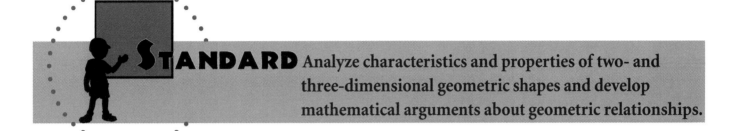

EXPECTATION: Recognize, name, build, draw, compare, and sort two- and three-dimensional shapes.

Explore

Younger students need to explore and sort pattern blocks. Discuss with them how the blue rhombus and the orange square are alike and how they're different. Students also need to see and examine shapes beyond the standard pattern blocks—squares in other sizes besides that of the standard orange pattern block and triangles other than the typical equilateral pattern block.

So pull out the building blocks and, in a class discussion, compare the shapes of the building blocks to the shapes of the pattern blocks: "This is a neat shape, boys and girls. It is flat on the top but this shape is what mathematicians call an arc. Where would a builder put this arc?" Encourage the children to work with the building blocks on their own, too.

Posters

NCTM recommends making sure students see a wide variety of triangles—ones with angles of different sizes and sides of different lengths. NCTM also notes that it's important for children to see shapes that closely resemble triangles but are not triangles, shapes that closely resemble squares but are not squares, and shapes that closely resemble rectangles but are not rectangles (NCTM 2000).

To accomplish this, start by creating 4 large posters. Title one of them "Triangles," one "Squares," one "Circles," and the last one "Rectangles." Encourage your class to search for these shapes in magazines and newspapers, and then cut out what they find and glue the examples to the appropriate posters.

When the posters are just about covered, ask your little mathematicians what they notice about the shapes. Be careful to limit discussion to one shape at a time. (This process may take quite a bit of time as thoughts snowball, so plan accordingly.) Your kindergarteners will note that triangles have 3 sides and 3 angles. (They'll want to call them "corners"; you'll want to tell them that the correct term is *angle*.)

Second graders will note that the sides are not always the same length. They may even want to measure to compare. Write their observations on the chart if there's

Standards are listed with the permission of the National Council of Teachers of Mathematics (NCTM). NCTM does not endorse the content or validity of these alignments.

79

Ask students to point out which are rectangles—and to explain their answers.

A QUICK TIP

For the sake of the children, you should purchase a couple of boxes of Toblerone chocolate. Toblerone chocolate comes in a wonderful triangular prism box. Do what you want with the candy, but let the children inspect the boxes and discuss their attributes.

room; if not, create another poster. The older children should also write their observations in their math journals.

Follow the same procedure for each other shape. Kindergarteners may observe that "all squares have 4 sides that are the same." Second graders may note, "Squares and rectangles have 4 right angles and 4 sides, but not all rectangles have equal sides like squares." Write down these observations on new posters.

As always, providing children with nonexamples can be a good way to help your little ones "get it." Draw shapes like the ones shown here and discuss with students what does or does not make each of these shapes a square or a rectangle or a triangle.

Geometry Museum

You'll need lots and lots of containers for your geometry lessons, so start by sending home a letter requesting them. If you like, you can use the reproducible on page 162 for this. You'll also want to add a few other objects, like a tennis ball (a perfect sphere), to supplement your discussions about the Geometry Museum.

As all those containers arrive in the classroom, place them in a part of the room labeled "Geometry Museum." Tell your students—and the custodians—that these objects are not junk, but rather artifacts for the Geometry Museum.

These artifacts can jumpstart many of your geometry discussions. Hold the cylindrical oatmeal box and ask, "How is this box like a soup can? How is it different? If I put this box on a slide, what would happen to it? Would it slide or roll? What shape is its face? How many faces does it have?" Follow with similar discussions about the pasta box or the cracker box or the cake-mix box.

As the Geometry Museum starts to take shape, read the story *Wayne's New Shape,* by Calvin Irons, about a chunk of wood that wants to change shape. With the class, discuss the changes the wood undergoes and the shapes that surround him on his travels. Be sure to use correct terminology: *sphere, cube, prism, pyramid,* etc.

- Hold up the tennis ball and ask, "How does this ball compare to Wayne?"

- Discuss the fact that both are spheres. They're different sizes, but both are spheres with no corners, no faces, and no edges.

- Hold up a cereal box and point out the faces and the edges on the box. Say, "How do the cereal box and the toothpaste box compare? They have the same number of faces and edges.

A Geometry Museum is a wonderful way to introduce geometric shapes. It helps reinforce the home-school connection, too.

Standards are listed with the permission of the National Council of Teachers of Mathematics (NCTM). NCTM does not endorse the content or validity of these alignments.

Let's count the flat sides. Those are called faces. There are six flat sides. This is a rectangular prism."

- Ask, "Can you find any other rectangular prisms in the collection from home?"

- Compare a cube and another rectangular prism. Ask how they are similar and how they are different. Explain, "This cube has six sides, so it is also a rectangular prism. However, it is a special rectangular prism because all of the faces are the same. Look very closely at the faces. What shape is this face, and this face, and this face?" (Keep turning the cube so all faces show.)

- "Right! All of the faces are squares. So this rectangular prism is called a *cube*. Can you find any other cubes in the collection?" Allow time for discussion and comparisons.

Tower Building

For this activity, your students will need the artifacts from the Geometry Museum as well as table space and/or floor space.

Begin by announcing, "I'm going to let you build with the boxes and containers that have arrived. You and a friend may try to build the tallest tower in the class, using these boxes from home. You may want to measure your tower once it is built. But here are the rules. You will need to share the boxes and containers with your classmates. You may not stand on a chair or desk or table. You must use quiet inside voices. When I say 'clean up,' you have to return all of the boxes to the Geometry Museum." This is one activity for which everyone will be on task.

Depending upon the ages of your students, you may want them to measure by simply "eyeballing" the structures and comparing the height of one to the height of another, or commenting on how tall the towers are compared to the height of one child: "Wow! This tower is as tall as Scott's shoulder!" or "Look! This tower comes to Robert's waist!" Last of all, have them measure with a meterstick or yardstick.

Allow children to build and create and have a great time. With each attempt at stacking, successful or failed, they are beginning to understand and internalize the properties of solid geometry.

INSTRUCTIONS FOR MATH JOURNALS

"Record what you have learned about building with geometric shapes. Then draw your tallest tower. If you can, label the shapes you used in that tower."

Even if a tower fails, the lesson doesn't!

Standards are listed with the permission of the National Council of Teachers of Mathematics (NCTM). NCTM does not endorse the content or validity of these alignments.

81

The Game of Shapes

Play a game. Say, "I see a shape with no edges and no faces. What shape do I see?" Allow the child who gets the correct answer (sphere) to give the next clues.

EXPECTATION: Describe attributes and parts of two- and three-dimensional shapes.

Geometry Slide

As a center activity, create a geometry slide for the solid shapes in your Geometry Museum by propping up one end of a piece of wood on a table. (A shelf from a bookshelf works well for this.) Also, make 3 simple signs out of index cards—one saying "Roll," one saying "Tumble," and one saying "Slide." Then proceed along these lines:

- Hold up the cards and review the meaning of each of the words with your students.

- Place those cards to the side of the slide.

- Hold a cylinder at the top of the slide.

- Ask the children, "If I were to let go of this cylinder at the top of this geometry slide, do you think it would slide, roll, or tumble down to the floor?"

- Allow discussion and predictions.

- Let the cylinder roll to the bottom. Tell the children, "It rolled. I'm going to put it over here by the 'Roll' sign."

- Invite the children to come to the slide in pairs or small groups during centers time and test the artifacts from the Geometry Museum—or other geometric solids—to see if they will slide, roll, or tumble. Tell them, "Once you have tested all the shapes, then please return the shapes to one spot so your classmates can do the same test. Pay close attention to what makes some shapes roll, some slide, and some tumble. Also, notice if any of the shapes can both slide and roll or slide and tumble."

INSTRUCTIONS FOR MATH JOURNALS

"Record in your math journal what you did. Tell which shapes slid, which ones rolled, and which ones tumbled, and why. You can draw this, write about it, or draw a T-chart to display your information."

These shapes roll
soup can
Pringles can
orange
These shapes slide
toothpaste box
cereal box
pizza box
These shapes tumble
parking cone
pyramid

Let the soup cans roll! Young mathematicians should always record their discoveries.

Standards are listed with the permission of the National Council of Teachers of Mathematics (NCTM). NCTM does not endorse the content or validity of these alignments.

Mystery Bag

For this activity, you'll need a mystery bag (a small fabric bag or pillowcase) with some real-world geometric shapes or geometric solids inside. Let's say one of those objects is a baseball or some other sphere. Proceed with steps like these:

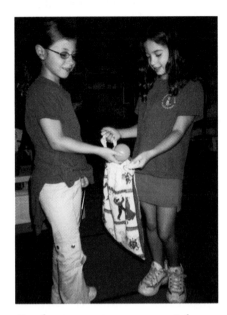

Students must use geometric language to describe the shapes in the mystery bag.

- Place the bag with its mysterious contents in front of the children.

- Pull out the baseball or other sphere.

- Make comments such as, "This baseball has no points, no faces, and no edges." (This is modeling geometric language.)

- Return the baseball or other sphere to the bag.

- Repeat for each other object in the bag.

- Ask one child to reach into the mystery bag, hold an object, and begin to describe the shape's attributes—all the while keeping the shape hidden in the bag. Encourage the child to use geometric language.

- Have children listen to the clues and identify the shape.

- The first child to give the correct answer is the next one to reach into the mystery bag and give new clues.

INSTRUCTIONS FOR MATH JOURNALS

"Write clues to describe the shape of something you would find in a grocery store."

> This shape has six faces. It does not roll. All the sides are congrewant
>
> What shape am I?

Another possibility for math journals: Have each student make up a riddle describing an item in the Geometry Museum.

Face Trace

- Ask children to describe the faces of a toothpaste box. Say, "Yes, each face is a rectangle. In fact, this face on the end is a square."

- Continue, "What are the faces on this Pringles can? Yes! There are 2 circles."

- Say, "Look carefully at the objects in the Geometry Museum. Choose an object and predict the shape of one of that object's faces. Then lay the shape on a piece of paper, trace around the face, and label it."

EXPECTATION: Investigate and predict the results of putting together and taking apart two- and three-dimensional shapes.

■ A Street Scene of 2-D Shapes

Assign a team of 3 to 4 children to create a streetscape of houses using squares, rectangles, and triangles of all sizes cut from wrapping paper, wallpaper, sandpaper, and construction paper. Either cut all paper ahead of time or let the children cut the shapes. Have students mount the shapes onto blue poster board to create a street scene and then write about their street.

INSTRUCTIONS FOR MATH JOURNALS

"In your math journal, describe or draw your street scene. Be sure to tell what shapes and sizes you used."

Here's an activity in which you actually encourage kids to "make a scene"!

The turkey's beak is actually a rhombus. Watch to see whether students understand that part of the directions.

Pigs & Turkeys

These two activities require reading and following geometric directions. The best part about the projects is that even though everyone follows directions (of course!), all the pigs and all the turkeys that result look very different, just as the children do! Is this reading or is this math? The answer is, "Yes!"

Let's begin with PIGometry. Before class, turn to the reproducible instructions on page 163 and make one copy for each student. In class, hand out the instructions. Students will also need glue, scissors, and a selection of black, pink, and gray paper. Ask them to follow all the directions. Celebrate the diversity that results!

For the turkeys, start with the reproducible on page 164. Proceed as above, with one exception: In this case, each student will need tan, gray, or brown paper, as well as a variety of other colors for feathers, beak, legs, and so on.

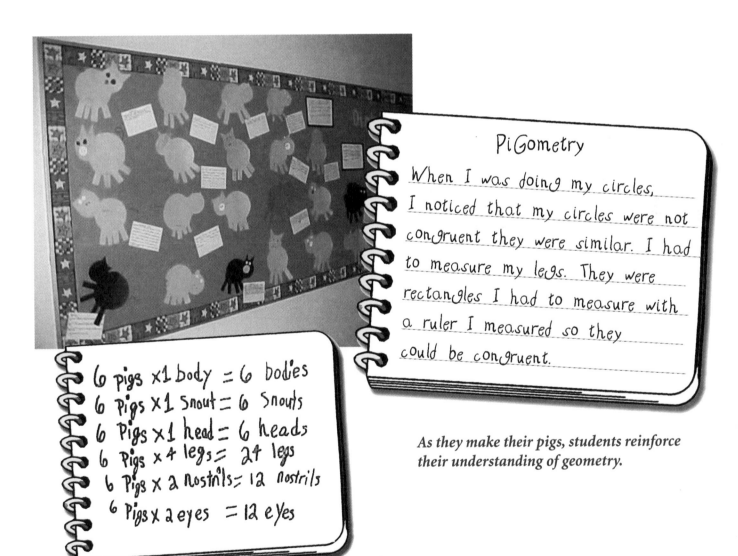

PiGometry

When I was doing my circles, I noticed that my circles were not congruent they were similar. I had to measure my legs. They were rectangles I had to measure with a ruler I measured so they could be congruent.

6 pigs x 1 body = 6 bodies
6 Pigs x 1 snout = 6 snouts
6 pigs x 1 head = 6 heads
6 Pigs x 4 legs = 24 legs
6 Pigs x 2 nostrils = 12 nostrils
6 Pigs x 2 eyes = 12 eyes

As they make their pigs, students reinforce their understanding of geometry.

Southwest Geometry

Read to your class the book *Coyote Steals the Blanket: A Ute Tale,* based on a folktale retold and illustrated by Janet Stevens. The story involves coyote's desert encounters with a hummingbird, a mountain goat, and a bighorn sheep. Unfortunately for coyote, he ignores the hummingbird's directions and steals a beautiful Native American woven blanket. What happens next explains why, to this day, coyotes are always racing around, never sitting still.

After reading the story, I implore my class to create a blanket for coyote so he can return the one he stole. This is the hook, and they always go for it!

To make the blanket, each child will need one 8-inch square of paper and four 4-inch squares. The large square should be gray. Each of the smaller squares should be either yellow, green, red, blue, black, or white—colors commonly found in Native American weaving. Tell the students, "You will need to select any 4 colors from these choices. Make sure you choose 4 *different* colors."

Children love the idea of creating a blanket something like this one for the coyote, and this activity helps them see the relationships among different shapes.

Ask each child to place her 4 small squares on her 1 large square. Talk about how all 4 squares cover the large square, so each of the small squares is one-fourth the size of the large square. Tell the children that the 4 squares are *congruent* to each other. Tell them that the 1 large square is *similar* to the smaller squares. Reinforce the concept that all squares are similar to all other squares.

Have each child fold and cut her 4 small squares in the following ways. She should end up with 2 large triangles, 4 rectangles, etc.

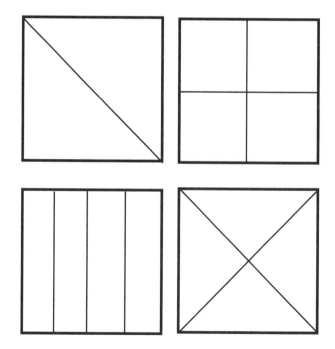

Guide the class in folding and cutting one square at a time. As appropriate for the grade level you're teaching, talk about the shapes that are created, the lines of symmetry, parallel lines, and angles.

Once the squares are cut, say, "Now you will need to take the pieces you have cut and arrange them on top of the gray square so that no gray is showing and no pieces are hanging over the edges."

Give students time to move and remove pieces. This is good practice for them. Once they've got all the pieces in place, have them glue their pieces to the gray paper.

Follow up by arranging the large squares into a class blanket. Cut some black paper into a fringe, add it to the outside edges, and you have a complete "blanket"!

ART TIP: Your finished class blanket will be more attractive if it includes a variety of color combinations. So even though each individual student will use just 4 colors, be sure to make all 6 available for students to choose from.

■ Baseball

A baseball field is filled with geometry. The infield is a square with 90 feet from home plate to first base, 90 feet from each base to the next, and 90 feet from third base to home plate. The bases are squares. Home plate is a pentagon. It's all geometry!

And that means that creating a baseball infield in miniature is a wonderful way to reinforce geometry concepts. Ask students, "Who wants to make a baseball field?" That one always gets an enthusiastic response! Continue, "If you listen carefully to me and follow directions, you will have a miniature paper baseball field. We're going to work on something called a scale. We'll make our fields measure 1 inch for every 1 foot on a major-league field." Then model how to make the baseball field.

- Start by cutting a 9-inch square. That's your field.

- Cut 1-inch squares for first, second, and third bases and glue them into position.

- Cut a pentagon for home plate and glue it in place. (Since pentagons are difficult for young children to measure, I simply ask each child to create a pentagon that is about the size of her 1-inch bases.)

If you like, take this project a step further by having children label the shapes and measurements. For instance, a child might write "9 inches" on the line from home plate to first base, and "1 x 1-inch square" on each of the bases. Challenge your charges to find as much math in their fields as they can and to label that math or write about it in their math journals.

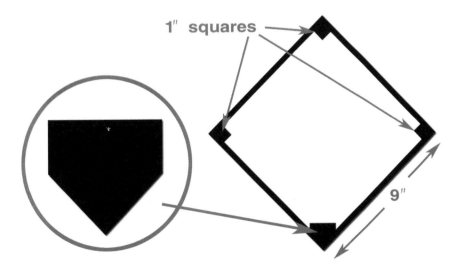

1″ squares

9″

A baseball infield is a square, each base is a square, and home plate is a pentagon. It's all geometry!

Two-Square Puzzles

Each child will need one 8-inch square of black paper, one 8-inch square of paper in another color, scissors, and glue.

Once everyone has the materials, model the folding and cutting for the children. Remind the class to *watch you* while you make each fold and cut; then they can follow your modeling one step at a time, folding and cutting their own papers. If they take their eyes off you, the project could be doomed! Here's the procedure:

- **STEP 1:** Compare the 2 squares, pointing out that they're congruent. "Look at these 2 squares. They're different colors, but they're the same size. Mathematicians call that *congruent*."

- **STEP 2:** Set aside the black square for now. Take the other square and fold it in half to form a rectangle. Ask, "What shape is this?" After students respond, continue, "When I open the paper, how many rectangles will I see?" (There will be 3—the 2 rectangles that formed with the fold and the 1 square you started with.)

- **STEP 3:** Go back to the rectangle and fold it in half to form a square. "What shape do I see now? That's right. Why is this new shape a square?" (It has 4 sides that are congruent and 4 right angles.)

- **STEP 4:** Open the square and ask, "How many rectangles do you see now?" (There will be 9, including the 5 squares.)

- **STEP 5:** Encourage discussion as the children describe the rectangles they see. Ask, "How many squares do you see now?" (There will be 4 smaller congruent squares and 1 larger square.)

- **STEP 6:** Cut apart the 4 squares that were formed by the folded paper. Set Square 1 aside for now. It stays whole.

- **STEP 7:** Take Square 2 and follow the same directions as for the larger square, creating four 2-inch squares. Cut those four 2-inch squares apart. Ask, "Do you see that we have 4 smaller congruent squares?" Add, "These squares are all similar to the 2 other sizes of squares."

- **STEP 8:** Pick up Square 3 and fold it by matching opposite right angles to create 2 triangles. Cut along the fold line so you have 2 triangular pieces. Ask, "What shape is this? Why is it a triangle?"

- **STEP 9:** Pick up Square 4. Fold it into a right triangle just like Square 3. Then fold it again into a second right triangle. This forms smaller triangles. Cut those 4 triangles apart. Ask, "The 4 triangles we just cut are similar to what other shapes that we've cut? That's right: the larger triangles."

- **STEP 10:** Now arrange the shapes on top of the black square to form a design. Don't let any of the shapes overlap or hang over the edges of the black square. Glue them in place.

Here's a colorful and very attractive variation: Pass out 8-inch squares in a variety of colors. After the initial folding and cutting, each child keeps one 4-inch square and trades colors with classmates for squares 2, 3, and 4. Then the folding and cutting continue as usual.

The four small orange triangles are congruent. The large yellow triangle is above the small blue square.

When students have finished, ask them to describe their designs in their math journals, using the correct geometric and directional words. As you review the journals, be sure to praise the use of geometric terms and positional language.

STANDARD

Specify locations and describe spatial relationships using coordinate geometry and other representational systems.

EXPECTATION: Describe, name, and interpret relative positions in space and apply ideas about relative position.

■ Coordinate Locator

When displaying student work, mount the pieces in a grid and label them with coordinates. As you refer to the pieces, of course you'll point out that the *x*-axis (the bottom, horizontal axis) is always named first and the *y*-axis (the far left, vertical axis) is always named second. A comma separates the two coordinates. Be sure that numbers or letters increase from left to right and bottom to top.

Now, instead of putting names on the front of papers, identify the author or artist with a list beside the bulletin board like this:

Amanda A, 1

Billy A, 2

Cassie A, 3

Dustin A, 4

"Can you find D, 2?" Engage your students by posting their artwork in a coordinate grid.

■ Back-to-Back Pattern Blocks

This game requires children to use correct geometric language and positional language. (Positional language appears in many states' language arts kindergarten standards, too.) You'll need to demonstrate a few times in the beginning, using yourself and a student.

- On the floor, sit back to back with a student. Make sure each of you has 1 pattern block in each shape.

- Decide who will be the leader and who will be the follower.

- The leader lays out a design using 5 or 6 pattern blocks that the other person can't see.

- Once the leader has completed the design, he gives directions that the follower must follow. The goal is for the follower to create a design that matches the leader's.

- The follower may stop the leader at any time and ask for clarification.

- Once all directions have been given and followed, both participants slide away and look at their designs. If the directions were given correctly and followed correctly, the designs should match.

- Rejoice!

FOR EXAMPLE

Maybe the leader says something like, "Pick up the orange square. Put that down first, close to you." The follower listens to the directions and does just that. Then the leader says, "Pick up the green triangle and place it above the orange square. Make sure the sides touch." Again, the follower does just that. The process continues until the design is complete.

Challenges: Once students understand how to do this and know the names of the shapes, they no longer use color names. Even more challenging for older students is to *describe* the shapes instead of naming them. For instance, instead of saying "Start with the square," the leader would say something like, "Start with the shape that has 4 sides all the same length and 4 right angles."

This activity reinforces understanding of geometric shapes. It teaches listening, too.

■ Stick It!

This is an easy activity that will give you a chance to assess students' understanding of plane geometric figures and of coordinate grids. Each child will need a copy of the reproducible grid from page 165, as well as pattern blocks in a variety of shapes.

If your students are not familiar with coordinate grids, you'll need to begin this activity by discussing the grid. "Boys and girls, take a good look at your papers. This is called a coordinate grid. Do you see the numbers on the bottom of the grid? Point to the numbers as I say them: 0, 1, 2, 3, 4, 5." Check that everyone is pointing to the bottom numbers. "What do you notice about the numbers on the left? Yes! There are the numbers 0, 1, 2, 3, 4, and 5 on the left also."

Continue, "Mathematicians have a special name for this grid. It's called a coordinate grid and it's a great way to locate information.

"Coordinate grids have numbers on the bottom. That line is called the *x*-axis. Do you notice the numbers begin with zero? Look carefully and you'll see that the numbers on the left begin with the *same* zero! The numbers on the left side, 0 through 5, are on the *y*-axis."

Test understanding of this new information by playfully saying, "Point to the numbers on the *x*-axis. Point to the numbers on the *y*-axis."

Continue, "When mathematicians want to tell you where something is on a coordinate grid, they say two numbers. The first number tells you where to *run* on the *x*-axis. The second number tells you where to *climb* on the *y*-axis. If I say, 'Put your finger on 2, 4,' you start at 0 and then run to the 2 on the *x*-axis, bottom row. Now it's time to climb up 4. So you climb up 4 lines and you land on the spot 2, 4 where the lines for 2 and 4 intersect." Repeat this several times to be sure your students can find each coordinate pair.

Once you're sure your students understand finding locations on a grid, say, "Find the triangle pattern block and place that shape on 5, 2." You can glance about your classroom to make sure that everyone has a triangle on 5, 2. This will also tell you if everyone knows the names of the shapes.

If your students are familiar with the shapes, challenge them with directions like, "Find the shape with 3 sides. Place that shape on 5, 2." For a greater challenge, say, "Find the shape with 3 congruent sides or 3 angles. Place that shape on 5, 2." Or instead of saying "Find the trapezoid," say, "Find the shape with 2 congruent nonparallel sides and 2 parallel sides." You can tailor your directions to the abilities of the students in your class.

Once they get the hang of it, ask the children to come up with the directions themselves. They love a chance to be the boss!

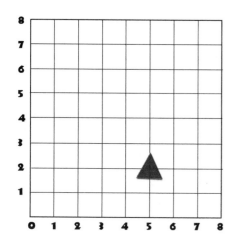

"Find the triangle pattern block and place that shape on 5, 2."

STANDARD
Apply transformations and use symmetry to analyze mathematical situations.

EXPECTATION: Recognize and apply slides, flips, and turns.

■ Slide It!

Slides are exactly what they sound like. A shape that slides to the left, to the right, up, or down remains the very same shape; it changes only its position. To demonstrate a slide, place a large cut-out of a 2-dimensional shape on the floor. Move it to the right and say, "Look at this square. Now watch as I *slide* it to the right. See? It still looks like the same square. It's just in another place. Watch as I slide it to the left. I can also slide it up and down. No matter where I slide it, it's still the same square." Repeat the same procedure using an arrow, a triangle, or any other shape. Give smaller shapes to each child and encourage students to manipulate the shapes to see how the shapes remain the same while only the position changes.

"These arrows show a slide to the right."

■ Flip It!

A shape that is flipped really flips. Shapes can flip on a vertical or horizontal axis. In either case, a flipped shape usually looks different. To demonstrate flips to your students, once again use a large 2-dimensional cut-out of a shape and show students what happens when you flip it. Then encourage them to do the same thing with their smaller cut-out shapes.

Demonstrate for students how some shapes—like squares, rectangles, circles, and equilateral triangles—can look exactly the same when they're flipped. The squares below show a flip, but the shape looks exactly the same.

"These arrows show a flip over a vertical line."

"These arrows show a flip over a horizontal line."

"When you flip a square, it doesn't look any different."

Turn It!

Shapes can also be turned. We describe turns in numbers of turns and also in angle measurements. A heart turned one turn is also turned 90 degrees. Once again, demonstrate with large cut-out shapes and then have students follow your lead with their smaller pieces.

"Here's a heart in its original position, and here's one turned 90 degrees."

Acrobatic Practice

Slides, flips, and turns can be demonstrated whenever your students are holding shapes. Take a few minutes and say, "Place your trapezoid on the table and slide it to the right. Flip it. Turn it one turn." Taking a few minutes every now and then to reinforce these concepts will help your students to retain them.

Pattern-Block Cloaks

Read *A Cloak for the Dreamer*, by Aileen Friedman. This story is about the creation of 3 cloaks by 3 brothers. Two of the cloaks are made of shapes that tessellate; those cloaks are successes. One cloak is full of gaps since the shape chosen (circle) does not tessellate.

Explain to students that tessellations are shapes that fit together and cover an area with no overlapping and no spaces between. Squares, hexagrams, rectangles, and equilateral triangles all tessellate. Circles do not tessellate since there are always gaps between the circles.

This book will most likely be an introduction to tessellations for your students. Once you've read it together, have them design their own "cloaks" with pattern blocks. Give each shild a poster -board cut-out of a cloak and explain that the child is to cover the cloak with pattern blocks. Or simply have each student cover an imaginary cloak on the tabletop or on the floor.

These students have created a cloak using rhombus and triangle pattern blocks. These blocks tessellate.

Symmetry in Print

Ask children, "Can you find examples of symmetry in magazine and newspaper pictures?" Either in class or as a fun homework assignment, have each student make a collage of pictures that demonstrate symmetry. Home, garden, and architecture magazines are especially good for this because they're often filled with examples of symmetry.

Symmetry Hunt

Go on a symmetry treasure hunt inside the school or outside on your campus. Bring the digital camera along and snap pictures of the symmetry your hunters locate. If you can, negotiate with the cafeteria staff to let your students explore the school kitchen. There is a great deal of geometry and symmetry there with ovens, racks, sinks, bowls, cookie sheets, lunch trays, and even the aprons!

Print the pictures you've taken and mount them on a chart. Ask the children to use crayons or markers to draw the "line of symmetry" on each picture. Be sure to label the chart—you might want to write something along the lines of "Symmetry in Our School" or "Can You See the Symmetry?"

Butterfly

Fold a large sheet of bulletin-board paper in half, making a sharp crease, and then unfold it. In front of the class, use pastel chalk to create half of a butterfly, drawing right along the fold. Refold the paper along the original fold line. Have kids remove their shoes and carefully walk over the paper in their socks, pressing lightly so that the pastel chalk transfers to the other side of the paper.

This activity generates lots of "oohs" and "aahs."

Unfold the paper. Wait. "Aaaahhhhhhhhhh!" will chorus from the class when they see the half butterfly become a full butterfly.

The chalk won't last forever, so let the kids paint over it. I let one child paint for 3 minutes, and then have the next child paint the symmetrical reflection for 3 minutes, and so on, until they have a complete butterfly. (This works well when you have a parent volunteer to control the one-at-a-time painters.) Hang the finished work where all can admire it.

Standards are listed with the permission of the National Council of Teachers of Mathematics (NCTM). NCTM does not endorse the content or validity of these alignments.

95

Blob 1

Children drop black paint blobs on white paper squares, fold each paper in half and press the sides together, and then open the paper. When the paint has dried, students cut each square on the line of symmetry. Leave these at a center so children can try to match the halves of the blobs.

INSTRUCTIONS FOR MATH JOURNALS

"Explain in your journal what you did to make your blob. Try to draw your design."

> I folded a paper in half, then I opened it up and dripped paint on the paper. Then I folded it again and pressed real hard. When I opened it up I had a beautiful blob of smeared colors. My blob was symmetrical and the fold was the line of symmetry.

Students should be able to describe their blobs in their math journals, tell how they created the blobs, and explain why they're symmetrical.

Blob 2

Each student folds a square of paper in half either diagonally or vertically. He opens the paper and drops paint colors on one side and then refolds it, presses the sides together, and opens it again. Once the paper is dry, the child draws the line of symmetry. Hang finished pictures on your bulletin board in a symmetrical arrangement. The total effect will be striking, I promise.

It's fun. It's artistic. It's a blob. And it's geometry!

Sandpaper Prints

This is a great project for art class. Using crayons, each student draws an object on a piece of sandpaper. Then it's time for you or an adult volunteer to step in. Place a piece of white construction paper over the design and press the white paper with an iron set on low. This transfers the picture to the paper. Mount the sandpaper and the white paper side by side so students can see that each is a symmetrical reflection of the other.

Kites

Read *Let's Fly a Kite,* by Stuart J. Murphy. Let each child use a construction-paper rhombus to create a symmetrical paper kite, then decorate the kite with paper cut-outs, stickers, or drawings, and tape a 24-inch piece of string to the bottom angle. Display these geometric masterpieces by hanging them from the ceiling or on a bulletin board.

Statue Maker

When you say, "Symmetry statues," the class gets very quiet. Of course they do. Statues don't talk! Then you add, "Can you make yourself symmetrical?" Each child strikes a symmetrical pose. (Giggles are permitted.) Next say, "Asymmetrical statues." Each child strikes an *a*symmetrical pose. This is a great time filler at the end of the day or for a few minutes before lunch.

Do students understand symmetry? They will if they've become symmetry statues.

Giggles are permitted as kids demonstrate asymmetry.

INSTRUCTIONS FOR MATH JOURNALS

"Draw your favorite symmetrical pose and your favorite asymmetrical pose and label which is which."

Pattern-Block Partners

For this favorite activity, students work in pairs or in groups of 4 to create symmetrical designs. The first child lays down 1 or 2 pattern blocks, arranging them in a pleasing design. Then the next student continues the design by adding a few more blocks in a symmetrical arrangement. The blocks can be added above, below, or alongside the other blocks. Count on your mathematicians calling you over to admire their beautiful creations.

Note: You might want to remind students that designs have limits and patterns go on and on. So when students create an arrangement with pattern blocks, it's a design, not a pattern.

A QUICK TIP

You can also get a geo-board for the overhead so you can demonstrate the use of the geo-bands for the class to see.

Geo-boards

In a whole-group setting, ask students to create symmetrical shapes using the math tools geo-bands and geo-boards. Once your students understand the concept and also understand that geo-bands are not rubber bands (even though they look exactly like rubber bands), place this equipment at a center and allow children time to create symmetrical designs on their geo-boards.

Architecture Center

This is a project parents can help with. Tell them you're collecting postcards of buildings and artwork that show geometry. Display the postcards at a center and ask children to write about the symmetry they see in the cards.

INSTRUCTIONS FOR MATH JOURNALS

"Choose one of the pictures in the Architecture Center. In your math journal, describe the shapes you see in the picture."

An Architecture Center gets parents involved and kids intrigued.

Students examine postcards in the center, and they learn to recognize geometric shapes in the world around them.

I am looking at the picture of Washington D.C. and I see squares in the windows, cylinders in the pillars and the doors look like really tall rectangles. I think there are triangles over the windows too.

"Leaf" It

Collect leaves in different sizes and shapes. Ask children to make crayon rubbings of the leaves, then mark a dark line along the line of symmetry in each rubbing.

EXPECTATION: Create mental images of geometric shapes using spatial memory and spatial visualization.

■ Fruit Salad

You'll need a variety of fruit that will become solid geometry for this lesson. Possibilities include cored pineapples and jellied cranberry sauce for cylinders; grapes, oranges, and blueberries for spheres; strawberries for cones; and lemons and limes for solid ovals. You'll also need a sharp knife for you to use and plastic knives for the students. In class, follow a strategy something like this:

- Ask children which 3-dimensional geometric shape each fruit looks like. (Example: The orange, which is shaped like a baseball, is a sphere.)

- Ask, "What shape will I make if I slice the cored pineapple parallel to the bottom? A circle? Let's see if you're right."

- Cut the slice and invite discussion of the resulting shape.

- Continue, still working with the pineapple: "What shape will I get if I cut the pineapple down the middle from top to bottom? Do you think I'll get a rectangle? Let's try it."

- Again cut in front of the children and discuss.

- Repeat similar predictions, cuts, and discussions for each other kind of fruit.

- Demonstrate how all slices cut from a sphere are circles.

- Allow children time to make their own cuts and discoveries with plastic knives.

Extension: Once all fruits are sliced, place some of the pieces at the art center and let children make fruit stamps by pressing the pieces into paint and then onto paper. Have them take this one step further by stamping patterns. (Patterns. Hmm. Sounds like algebra to me.)

ART TIP: Did you know that a halved orange dipped into orange paint can become a perfect mini-pumpkin? A halved lemon pressed into brown paint can become a football; just add white chalk laces.

A QUICK TIP

How is this math? We want children to turn language into image. When they look closely at a lemon and picture the shapes *inside* the lemon, they're turning language into image.

Ask students to find all the shapes the artist used.

■ Make an Art Connection

Show an art print by a famous Impressionist or Cubist such as Cézanne or Picasso and ask children to name the geometric shapes that the artist painted.

■ "Shapes" Poem

You'll need a copy of Shel Silverstein's poem "Shapes" from the book *A Light in the Attic.* You'll also need a sheet of white drawing paper for each child. In class, pass out the drawing paper and then proceed like this:

● Explain, "I'm going to read this poem three times: the first time just for listening, the second time for drawing, and the third time for checking. You'll need to listen very carefully each time."

● Ask, "Ready?"

● Read "Shapes."

● Say, "Now you may draw what you heard in the poem. Do that as I read it again."

● Read the poem a second time as children draw their pictures.

● Say, "Now I'll read it one more time while you check your drawings."

● Read the poem once more, then show Shel Silverstein's illustration.

● Ask, "How does his picture compare to yours?"

● Children draw new pictures or color the sketches they drew.

EXPECTATION: **Recognize and represent shapes from different perspectives.**

■ Ribbon Shapes

Start with a strand of ribbon 8 feet long. Tie the ends of the ribbon together and then proceed along these lines:

● Ask 4 children to come to the front of the class and hold on to the ribbon.

● Say, "When I name a shape, you 4 children are to move the ribbon into the shape I name. No talking, but you may look at one another and see if you can communicate with your eyes or your hands."

● "Ready? Square!"

● Children will laugh, roll their eyes, and form a square.

● Repeat for other shapes.

How quickly can your students form the shape you name?

Village of Round & Square Houses

This activity is based on the book *The Village of Round and Square Houses,* written and illustrated by Ann Grifalconi. The exquisitely illustrated narrative takes place in the real-life village of Tos, which lies in the remote hills of Cameroon in central Africa. In this tale, a grandmother explains to her granddaughter how, after running from the town during a long night of tremendous noise and flowing lava, the people returned to their village to see only 2 houses remaining: 1 round and 1 square. Since then, in Tos all houses for women are round and all houses for men are square.

Divide the class into 2 groups: one of all the girls and one of all the boys. Call each group separately to an area on the floor where they can be close to you. Tell them to bring scissors, rulers, and pencils. Give directions one step at a time, modeling each step as you go.

To complete this project, each child will need:

- Scraps of green construction paper
- Scissors
- Ruler
- Glue
- Pencil

Each girl will also need:

- 1 piece (4 x 18 inches) of tan construction paper
- 1 piece (9 x 9 inches) of brown construction paper
- 1 carpet square for a work mat
- Hole punch for the compass
- 1 strip (1 x 5 inches) of index card for the compass
- 1 thumbtack for part of a compass

Each boy will also need:

- 1 piece (4 x 17 inches) brown construction paper
- 1 piece (6 x 12 inches) of tan construction paper

This is a challenging activity, but if you model one step at a time, the girls should end up with houses something like this one.

Round homes for the girls:

- Say to the girls, "I'm laying my large tan paper rectangle out so that it's a *wide* rectangle. I'm going to find where I think the middle of this rectangle is, so I know where to put a door."

- Demonstrate by gently matching right angles and pressing ever so lightly at the middle, making a slight fold at the top of the paper. Say, "Girls, this fold is going to be very light, but you are so smart, I know that you'll be able to see it. We'll use that fold mark to center the door.

- "Now I'm going to go to the bottom of my rectangle. I'm going to start at that middle point and I'm going to measure a 2-inch vertical line."

- Demonstrate.

- "Now I'm going to start at the top of my 2-inch vertical line and I'm going to measure a 1-inch horizontal line."

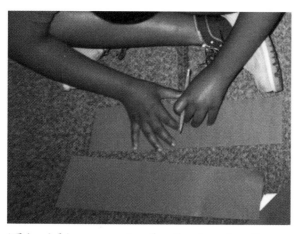

This girl is measuring for the door in what will become the cylindrical base of her house.

- Demonstrate.

- "Now I'm going to cut along both those lines to make a door."

- Demonstrate.

- "Now I'm going to fold the cut paper to the outside to resemble an open door."

- Demonstrate.

- "Now, make that whole sheet of paper into a 4-inch-tall cylinder by gluing the short ends together." (I do not tell them how much paper to glue. Some will choose to make wider cylinder homes than others. This independence and individuality is very appropriate for young children, and I encourage it!)

- Next, model how each of the girls should make her brown square of paper into a cone-shaped roof. To do this, start with a simple homemade compass. Place the brown 9 x 9-inch paper on top of the carpet square. Use the hole punch to make a hole in one end of the index-card strip. Push the thumbtack through the opposite end of the index-card strip and then through the center of the brown 9 x 9-inch paper (still working on top of the carpet square). Slide the pencil through the punched hole and draw with a circular motion. This will result in a perfect circle approximately 9 inches in diameter. (It will also reinforce the concept that all points in a circle are equidistant from the center of the circle.) Cut out the circle.

- Make a cut from any point on the outside edge of the circle directly to the thumbtack hole.

- Overlap the edges of the paper, and then glue the edges together to create a cone.

- Cut leaf shapes from the green construction paper and glue them in place to decorate the roof.

- Place the cone roof on top of the cylinder.

Using a homemade compass demonstrates that all points in a circle are the same distance from the center. Creating a roof from that circle shows the relationship between a circle and a cone.

Square homes for the boys:

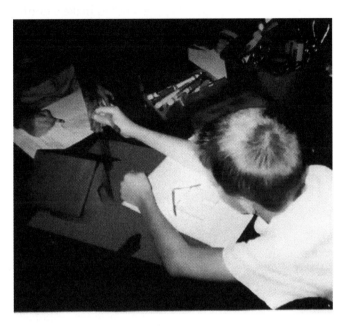

This boy has finished his square house. Now he's measuring it and recording his findings in his math journal.

- Say to the boys, "I'm going to take this 4 x 17-inch paper and I'm going to start at the 4-inch side. I'm going to fold back 1 inch of the paper to create a tab. Watch me measure one inch."

- Slowly demonstrate measuring and folding.

- Say, "We're going to fold the remaining 16 inches into exact fourths." Challenge them to tell you how to do that. (Measuring will not be necessary as folding will result in halves. If one of your little smarty-pants wants to measure, explain, "We won't need to measure because folding in exact halves will result in fourths.")

- Fold the 16 inches in half widthwise. (Do not include the 1-inch tab that was created.)

- Next, fold each of the halves in half. (This is widthwise again.) Without any measuring, these folds will create 4 equal-length walls.

- Fold the paper so that it forms a square when placed on end. Glue the 1-inch tab to the adjacent wall.

- To begin to make a rectangular door, start at the base of one of the walls. Measure and then cut along a 2-inch vertical line from there.

- Make a 1-inch-long horizontal cut starting from the top of the vertical one.

- Fold the cut paper to the outside to make the cut piece look like a door.

- Fold the tan paper to make a tent-like roof.

- Add green leaf cutouts to the roof for decoration.

- Place the roof on top of the house.

INSTRUCTIONS FOR MATH JOURNALS

"Write about the shape of the house you made as well as the attributes of that shape. Include an illustration with labels. You may also wish to measure the house you made and record the width, height, perimeter, and/or circumference."

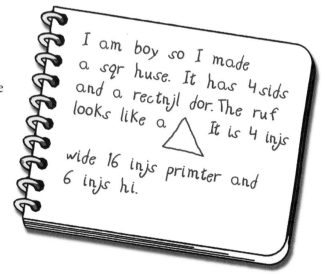

I am boy so I made a sqr huse. It has 4 sids and a rectnjl dor. The ruf looks like a △ It is 4 injs wide 16 injs primter and 6 injs hi.

EXPECTATION: Relate ideas in geometry to ideas in number and measurement.

■ Rockets

Ask parents to send in cardboard tubes from paper towels, toilet paper, and gift wrap. (This works especially well around the holidays.) Put the tubes at the art center and let each child paint a tube. Later put out paper cut into specific shapes as well as odd scraps of paper. Invite children to turn those into geometric designs on the rockets. If they like, children can attach toilet-paper tubes for fuel tanks.

Encourage each student to create a nose cone for her rocket by cutting paper into a circle, cutting once from the outside edge to the midpoint, and then fashioning the cut paper into a cone. Once rockets are complete, children can add tissue paper to represent the exhaust. Of course, each child should measure the length, circumference, and diameter of her spaceship.

INSTRUCTIONS FOR MATH JOURNALS

"Write about the shapes in your rocket. Be sure to include the rocket's length, circumference, and diameter."

A QUICK TIP

Rocket into this activity by reading *Shape Race in Outer Space,* by Calvin Irons, and/or *Captain Invincible and the Space Shapes,* by Stuart J. Murphy.

On a cardboard rocket, a paper circle becomes a nose cone.

■ Cheez-It Shapes

Cheez-It crackers are pretty close to 1 square inch! That means they can be used for some really neat math activities. Here's one of them.

● Pass out 5, 6, 7, or 8 Cheez-Its and a piece of drawing paper to each child.

● Say, "What geometric shape does 1 cracker remind you of?"

● "That's right, a square!" "Why?"

● "You got it! It has 4 sides, all sides are the same (they're congruent), and it has 4 right angles."

- "Is it anything else?"

- "Yes, it is a polygon, since it has 3 or more sides and it is a closed shape. Yes, it is also a rectangle since it has 4 sides and all 4 angles are right angles!"

- Add, "I want you to arrange the crackers in any shape that has every side of every cracker touching another cracker and no crackers overlapping. Then, trace around your new shape. See how many shapes you can create using __ crackers. Do you think you can create rectangles? Squares? Triangles? Circles?"

- Turn them loose and let them create, draw, and count.

When a student creates a rectangle with 6 crackers, she'll see that 3 sets of 2 are 6. She'll see that 2 sets of 3 are 6. So are 6 sets of 1 and 1 set of 6. They're all 6! That's number sense!

EXPECTATION: Recognize geometric shapes and structures in the environment and specify their location.

■ Geometry Hunt

Because I want my students to see geometry everywhere, even in their homes, I ask them to complete a geometry survey. This connects the classroom to the real world. You can do the same by copying the Geometry Hunt reproducible on page 166 and sending it home as homework. Explain to students that they may use words and/or pictures to document the geometry they find at home. (I find this activity is an especially good one because there's little chance students will complete it without conversing with their family using their newly acquired mathematical language.)

■ Flags from Around the World

Most flags are made up of geometric shapes. If your media center has reference books showing flags from other countries, those flags can generate discussion on shapes and provide the children with a social-studies lesson at the same time. Thanks to the Internet, you can also print out flags from all over the world. Just be sure to respect copyright protections. (If your school has an Enchantedlearning.com membership, you may want to use that site, which includes many printable flags from around the world.) As the children look at the flags in the reference books or as printouts, ask them to name the geometry in each flag.

Oh, say, can you see all the geometry?

105

MEASUREMENT

The NCTM standards for this strand encompass a great deal: length, volume, weight, time, and area. Many states also include money. These are all very abstract concepts. Marilyn Burns has noted that the way children develop proficient measuring skills is through lots of firsthand practice (Burns 2000). Looking at pictures on a workbook page will not help your students understand and use measurement.

Instead, they need to hear, ponder, and answer questions that compare measures. Who has a foot that's shorter than yours? Who has a foot that's longer than yours? Who has the shortest foot? Who has the longest foot? Who is taller than you? Who is the tallest in your class? What is lighter? What is heaviest?

That doesn't mean the answers will necessarily be precise. Children should also be estimating. "That bunny looks very close to the size of my kitten." "My dad is about that tall." "This cookie weighs about as much as that paper plate." "These 2 apples weigh about the same." We want our students to be able to come up with real comparisons. That takes practice.

The key to all of this is that it helps students see how this strand connects to their world. And you can take it further. Brainstorm with your class. Discuss what your classroom would look like if the builder didn't measure anything, what the food would taste like if the cafeteria ladies didn't measure the ingredients for the cookies, or what would happen if a tailor who sews beautiful suits didn't measure first. Yikes! Measuring is very important!

It's also pretty challenging. After all, in the primary grades children are learning not only *how* to measure, but also *what* to measure. Let's see, can I weigh this box to determine how wide it is? Can I use a ruler to figure out how much water is in this cup?

This strand is challenging for another reason, too. Many of your students come from homes where the mom or the dad makes fuzzy math comments like, "I'll read to you in *just a minute*. Let me call your grandma, bathe your baby brother, and put the dishes away. Just wait a minute!" Or after a few months away from the grandchildren, Grandma declares, "You must have grown a foot!"

We know why this is one tough strand to teach!

Of course, to make things just a bit more complicated, students in kindergarten and in first and second grades are using both nonstandard and standard forms of measure. This is a lot of work for us and for our students. But we can do it, and we can make the learning engaging and interesting so they'll remember the lessons. That's what this chapter is all about.

It's not easy teaching measurement! Be sure to give your students lots and lots of opportunities to practice.

STANDARD
Understand measurable attributes of objects and the units, systems, and processes of measurement.

EXPECTATION: Recognize the attributes of length, volume, weight, area, and time.

How Long Is It?

Say, "I can measure the length of this book." Then demonstrate the possible ways to measure. In kindergarten and early first grade, you can cut string that equals the length, lay paper clips along the length of the book, or lay the book on paper and then draw a line on the paper showing the length. Students then compare which string is longer, which line is longer, and so on.

Next, pose this question to the class: "What other things can we measure for length?" When they've decided, ask each child to measure whatever it is that he's chosen. He can draw a line that shows how long his foot is, one that shows how long his finger is, and others that show the length of whatever else he chooses. Let your class cut twine, string, or yarn to show the lengths of the items they've selected.

How Much Will It Hold?

In the early grades children need to spend time pouring rice, beans, and water from one container into another. It is important that containers vary in size and width. The size of the openings of the containers should also vary. Children are often surprised when they discover that the short, wide container holds more than the tall, thin container. Ask them to use comparison language: "This bowl holds more rice than this pitcher." "The cup holds fewer beans than the jar."

Splash! Who Spilled the Water?

Read *Mr. Archimedes' Bath,* by Pamela Allen. Kids love this book. Be prepared for the page with a man about to step into the bath. The reader views him from the back and—gasp!—he's not wearing clothes (which is one of the reasons kids love this book)! Each time another animal gets into the tub, no matter what the size of the animal, water from the tub spills out.

Use this book to generate discussion: Why is the water spilling out? What could possibly be making the water spill out? The story is engaging and kids are enchanted with the characters. (This is also displacement of water.)

Standards are listed with the permission of the National Council of Teachers of Mathematics (NCTM). NCTM does not endorse the content or validity of these alignments.

109

Variations: Pamela Allen's *Who Sank the Boat?* is another great story on a similar concept. So is *Room for Ripley,* by Stuart J. Murphy. In that one, Ripley the Guppy comes home from the pet store and his new owner uses cups, pints, half-gallons, and gallons to fill up Ripley's simple goldfish bowl. As you read the story to your class, pour water into an empty dishpan or terrarium, using the appropriate measuring implements.

Weigh Everything You Can

Children must have experiences weighing and otherwise comparing objects. Looking at pictures in a book cannot begin to help a child understand that the rock weighs more than the feather or the scissors. She needs to hold items in her hands and compare their weights. She needs to pour, fill, and refill containers. She needs to balance items on a scale. And she needs to measure lengths of different objects and compare them. *That's* how a child internalizes these concepts.

Hummingbirds & Panda Bears

Teachers of kindergarteners and first graders need to point out attributes whenever they can. Say things like, "Wow! That backpack is full! It's very heavy." "The estimate jar is full." "This box may be large, but it's very light." "The hummingbird is very light. The panda bear is very heavy."

How Much Paper Can 8 Cheez-Its Cover?

Much as you did in the geometry chapter, ask, "How many different arrangements can you create using just 8 Cheez-Its?" Instruct each child to lay 8 crackers carefully on a piece of paper and then trace around them. Encourage children to create a variety of shapes using only 8 crackers. A child might make a rectangle of 2 x 4 or 4 x 2, an L, a U, and so on. This is a good way for students to see that 8 crackers can be arranged in a variety of ways—and that in turn builds the foundation for discussing the concept of area. Children need to know that an area of 8 can come in a variety of shapes.

Hand out more of these mini-measurers and ask, "Can you create arrays for 12 using 12 Mini-Wheats or 12 Cheez-Its? Last time we made any kind of shapes. We made U shapes, L shapes, even stair steps. Today let's make just arrays. Do you remember that arrays are rectangles and that the number in each row is exactly the same?" After students create these arrays for 12, ask them to make more arrays for other numbers.

LESSONS TO MUNCH ON

Did you know that each regular-size Cheez-It is about 1 square inch? (You do if you've already read the geometry chapter!) This makes them wonderful tools for exploring the concepts of area and measurement.

QUESTIONS OF TIME

For younger children, you need to reinforce the days of the week—as well as the concepts of yesterday, today, and tomorrow—each day at calendar time. As students gain knowledge, talk about the months and seasons, too. Challenge older children to look at the date and ask, "What will the date be in two weeks? Two weeks and three days? One week from yesterday? Two weeks from tomorrow?"

 ## Time Talk

In your classroom, make constant references to time. Say, "We have lunch in 15 minutes. The clock will look like this [draw a quick sketch on the board or move the hands on the class Judy clock]." "The assembly is at 10:30. That's 2 hours from now." "I hope to meet with the Panther reading group in 10 minutes, at 9:00."

And then there's the biggie: "I'm giving you 5 minutes until cleanup." Now this is a hard one for teachers to stick to! But you must. If you say, "5 minutes until cleanup," be true to your word. If you let 5 minutes drift off into 10 because the students are so into their work, you'll confuse them.

Instead, at 5 minutes say, "I told you we'd clean up in 5 minutes. As I look around I can see that you are still working, so I'm going to give you 5 more minutes." That way they understand 5 minutes and you can let them finish.

The same thing goes for "10 minutes more." If you look up after 3 minutes and suddenly bedlam is ensuing because they've finished, say, "I know I told you 10 minutes more, but I can see that you're all done. So even though it's only been 3 minutes, let's clean up. Good for you!"

 ## Read About Time

Read *The Very Hungry Caterpillar,* by Eric Carle. Discuss the time lapse from page to page and the time lapse from the beginning of the story to the end.

Then read *The Grouchy Ladybug,* also by Eric Carle. As you read, move the hands on the class Judy clock to match the time each new character enters the story. If your students have individual Judy clocks or paper clocks, they should move the hands on their individual clocks at the same time. Discuss the passage of time from page to page and from the beginning of the book to the end.

Read Pat Hutchins's book *Clocks and More Clocks.* Ask children to move the hands on their clocks while you move the ones on the class Judy clock to match the times in the story. Discuss what is going wrong in this story and why the clocks don't match. After a few pages your "quick studies" will figure out that the clocks are all correct. It's just that as you move from room to room, the time moves on, so it *looks* like the clocks don't match.

Read *Get Up and Go!,* by Stuart J. Murphy, to the class. Then read the book a second time, emphasizing the timeline that continues on each page. Together with your class, create a timeline of your day in school. Ask each child to create a timeline for her mornings, evenings, or weekends. Whether the timelines are by the hour or by the minute will depend upon the ability of your class. This is real-world math at its best!

Ladybugs & Others

Compare the lengths of the animals and the sizes of the fonts in *The Grouchy Ladybug,* by Eric Carle. (The font and the animals get larger with each spread in this book.) Let the children use string, Unifix cubes, centimeter cubes, or centimeter rulers to compare critter lengths and font sizes. Place the book on the table and model for students the method you are using to compare the sizes of the animals. Tell the children to measure each creature at its widest or tallest point. (With the exception of the snake, which happens to be wrapped around a tree, the animals are fairly easy to measure.) Older children should record the length of each animal in chart form. If you like, use copies of the reproducible on page 167.)

Once again, pull out the math journals!

The grouchy ladybug

1. y.J. about 5 c.m. from the top to the botom

2. Stag Beetlet 9 c.m. the top to the botom of the St. be.

3. Praying mantias 13 c.m. from the top to the botom

I nodest that 5+4= 9 9+4=13 and every time he met a insx the bigger it got

Weighted Comparisons

Does a plastic sandwich bag of pennies weigh more or less than a bag of nickels or a bag of quarters? Does a bag of nuts and bolts weigh more than a bag of small screws or a bag of other small hardware? Let children compare with pan balances. And let them think of other items to put in clear plastic sandwich bags and compare. Vary the bag sizes and vary the number of items in the bags—you want your little mathematicians to learn all they can!

How Long Is That Carrot?

Put out several carrots or several ears of corn and invite younger children to arrange the vegetables from longest to shortest. Ask first and second graders to measure each vegetable to the closest inch or half-inch.

Then set out a variety of fruits and veggies—you might include an apple, a potato, a squash, an ear of corn, and a carrot stick—along with a balance or pan scale. Ask the children to weigh each piece and line up the items from lightest to heaviest.

DOLLAR SAVERS

Make friends with the employees in your grocery store's produce department. Tell the produce manager you're happy to take, for a super discount or for free, any produce he's about to throw out. Explain that you're not going to eat the fruits and veggies; you'll just use them to help children learn. Be sure to send a thank-you note from you and your class.

Baseball or Baking

Select a theme, such as sports, and place sports equipment (a hockey puck, a hockey stick, a softball, a batting glove, a baseball) at a center to measure and weigh. Call on your students to help you think of other themes and the corresponding items to measure and weigh. Imagine a "baking center" where children can work with measuring cups, wooden spoons, cupcake tins, spatulas, aprons, and measuring spoons!

A Is for Apple

A QUICK TIP

Weighing apples is a great activity for celebrating Johnny Appleseed's birthday in late September.

Ask each child to bring an apple to school—or bring them yourself and give an apple to each child. Ask children to predict who has the heaviest and lightest apple at each table and then prove their predictions with pan or bucket balance scales.

Repeat this procedure with Beanie Babies, rocks, pinecones, or anything else that will fit into the scale.

Bug Parade

Read *The Best Bug Parade*, by Stuart J. Murphy, with the class. This book emphasizes comparative language by using terms like big, bigger, and biggest. It's a great lead-in for discussing these terms with your students.

The Kissing Hand

Read Audrey Penn's *The Kissing Hand* with the class. Then ask children, "How many kisses can you fit in your hand?" After bewilderment sets in explain, "I'm talking about Hershey's Kisses. How many of those can you fit in your hand?" Then continue with something along these lines:

"*How many kisses can you fit in your hand?*"

- Each child traces his hand.

- You give each child 2—and *only* 2—wrapped kisses.

- You ask, "How can you figure out how many kisses you can fit in one hand when you have only 2 kisses to start with?"

- The students problem-solve.

- Once each child has determined a reasonable answer, he goes to the "Kisses Center," fills his hand with kisses, and counts to see if he was correct. Of course, he can't *eat* all the kisses since his friends need kisses to see if they were correct also.

- Homework that night is for each child to trace his mom's hand and predict how many kisses will fit in that larger hand.

- The next day children return to the Kisses Center, this time with the hand tracings and predictions, to cover their tracings with kisses and see if their predictions were correct.

- Once all predictions are checked, the real problem sets in: What on earth do you do with all the extra Hershey's Kisses? Ask for suggestions. Let the children really think about this one. They'll want to eat the candy, so they'll be invested in this discussion.

- If there are no clear favorite ways to solve the problem, use repeated subtraction. You might say, "Okay, everyone take 2 kisses. Are there more kisses left? Then take 1 more. Still more kisses left? You can take 2 more kisses." This is a reasonably fair approach and it shows that division is really repeated subtraction.

Area of a Hand

Ask, "When you spread your fingers wide apart, is the area of your hand greater than, less than, or equal to the area of your hand with your fingers together?" Children do not get this one until they prove it to themselves. Really. Let each child trace her hand with fingers together and then trace it again with fingers wide apart. Suggest the child find and count how many of one thing can fit in each tracing and then compare. Small items like cereal pieces, centimeter cubes, or mini-marshmallows work well for this.

About How Far Is a Mile?

Adults often refer to distance from one place to another place in minutes. Home is "15 minutes from Grandma's" instead of 10 miles. The grocery store is "5 minutes away" instead of 3 miles. As a result our children don't have a clear understanding of miles. We need to remedy that!

With the class, read *Henry Hikes to Fitchburg*, by D.B. Johnson. The story is really about Henry David Thoreau, but in this case Henry is a lovable bear. As Henry goes on his journey, the book records the miles he travels.

After reading this story, create a "Hiking from Our School" bulletin board. Ask a child to draw your school, then ask others to draw landmarks that the children know—they might show the local fast-food restaurant, the movie theater, the grocery store, and/or the park. Have students paint or draw a path leading from the drawing of the school to the drawings of the various town landmarks. As you provide the information, ask them to label (in miles) the distances between locations.

Pencils, Clips, Pennies & Clothespins

Children need to understand that the desk that is 43 pennies long will not be 43 clothespins long. They will realize this only after you give them many opportunities to lay items one after another from one end of the desk to the other end. So give them those opportunities. They must also understand that when they're measuring with nonstandard units, all units must be the same length. Be sure to help them discover that. Not all crayons are the same length, especially after a 5-year-old has chewed a few!

Don't be surprised if some children don't start at the very end of the item to be measured or don't lay the paper clips, pennies, or crayons so they're touching end to end. Be sure to correct them because these will be the same children who will measure the desk starting at 3 on the ruler or who will start to measure at zero but place the ruler an inch or 2 from the end of the desk.

Cereal & Cracker Rulers

Create nonstandard rulers with Mini-Wheats or Cheez-Its. Cut pieces of poster or foam board 1 x 12 inches and give one to each child. The child glues 12 pieces to each strip to create a "ruler." Ask, "How many items can you find in the room that are *exactly* 1 Mini-Wheat or Cheez-It long?" "How many that are 2 Mini-Wheats or Cheez-Its long?" Continue through 12.

Each Cheez-It is 1 inch square, so 12 of them make a fun ruler.

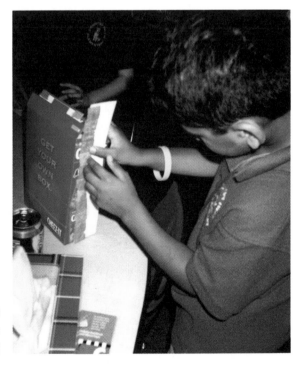

Invite students to measure everything in sight with their Cheez-It rulers!

Inchworms

Read *Inch by Inch*, by Leo Lionni, with your class. Create 1-inch inchworms by gluing tiny pom-poms to 1-inch squares and then gluing each square to the end of a Popsicle stick. Students can measure items in the room using their own inchworms. (These inchworms can also help children space words when they're writing.)

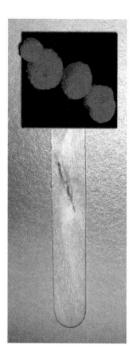

Popsicle-stick "inchworms" make measuring practice fun.

~~~~~~

EXPECTATION: Select an appropriate unit and tool for the attribute being measured.

~~~~~~

Which Weighs More?

When introducing any form of measurement, explain what attribute you're measuring and what tool you're using. Also review how to use that tool. "I'm going to weigh this beanbag toy and see if it weighs more than this bag of marbles. So I'm going to use the balance scale. It tells me what weighs more. Watch as I place the bag of marbles on one side and the beanbag on the other. Do you see how the side with the bag of marbles drops closer to the table? That means that it's heavier than the beanbag." Make sure your language is specific.

Or, for those who are using standard measurement, say, "I want to know exactly how much this beanbag bear weighs. I'm going to use the balance scale. Watch as I put the bear in one bucket and then place the grams [or "gram stackers"] one by one in the other bucket. As I put the grams in, I'm going to count them and I'm going to watch until the buckets are even. When they're even, that will tell me that both buckets weigh the same. Who can guess what that means about how much the bear weighs?"

How Long Is Your Shoe?

Model how to measure length using standardized measuring tools: "I'm going to measure this shoe to see how long it is. I'm going to use the ruler. I'll put the end of the shoe at the end of the ruler. I'll make sure the ruler is at zero. How many inches long is the shoe?" Then have students follow your example.

String Around a Bear

Explain to the class, "I want to know how long this stuffed bear is, so I'm going to use the meterstick." Or "I wonder how wide around this bear is. Since I can't measure his tummy with a flat ruler, I'm going to use a string. I'll wrap the string around his tummy and cut the string at the point where it meets. Then I'll measure the string against a ruler or yardstick [or a meterstick if you are using metrics]." Demonstrate as you talk.

STANDARD

Apply appropriate techniques, tools, and formulas to determine measurements.

EXPECTATION: Measure with multiple copies of units of the same size, such as paper clips laid end to end.

Super Sand Castles

Read *Super Sand Castle Saturday,* by Stuart J. Murphy. This story is a perfect bridge from standard to nonstandard measure. Working with your students, paint a mural of sand castles, snowmen, or anything else your class desires. Once the paint is dry, let children drop a glue line alongside each object and then glue a series of identical items to the glue line. They might use buttons, mustard packets, Band-Aids, toothpicks, straws, or whatever you can dig up from the craft closet. Count the items and label the mural with words like, "This snowman is 12 ketchup packets, 8 feathers, or 35 inches high."

ART TIP: If you want to paint snow to look real, mix equal parts liquid Elmer's glue and shaving cream and then apply the "paint" with a brush. Let it dry for 24 hours. It dries looking like real, honest-to-goodness snow!

This is an ideal bridge between nonstandard and standard measure.

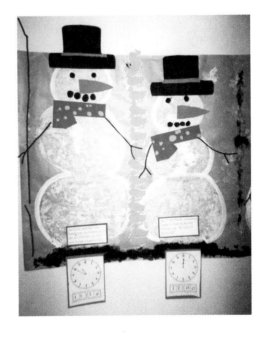

This mural has a story to tell. The labels say that at 11:50 the snowman was 4 pipe cleaners tall, but he was sitting in the hot Florida sun. By 12:00 he was 15 ducks tall.

EXPECTATION: Use repetition of a single unit to measure something larger than the unit, for instance, measuring the length of a room with a single meterstick.

How Many Shoes Are You?

Ask each child to take off one shoe. (This is already a hit— they love to take off their shoes!) Say, "You're going to measure a friend and a friend is going to measure you. We're going to find out how many shoes tall you are." Demonstrate with one child lying on the floor. Place one of that child's shoes at the child's heels. Put your finger at the top of the shoe, then move the shoe so it's immediately above your finger. Continue until you've measured the full length of the child. Say, "You are 5 [or whatever you've measured] shoes high." Have all students pair off and repeat the process. Then compare results and discuss: "Do most children have the same ratio between their height and their shoe size? Is it typical for 5 shoes to equal 1 Carol?"

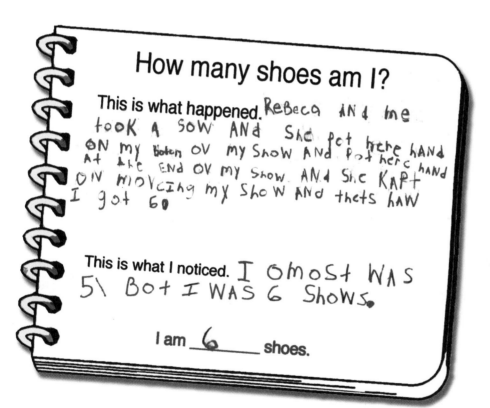

It's shoe time! Measuring with shoes gives kids plenty to write about.

Pick a Tool

Try this on your little geniuses. Set out a stuffed bear, a balance scale, a measuring tape, a ruler, and string. Ask them to find the length of the bear and its weight. Watch to see if they understand which tool to use. Follow this procedure each time you want them to measure. When you set out just the measuring tape, you're already telling them what tool to use. You've taken away part of the math! So when measuring, let them select the tool and take it from there.

Hand or Foot (Grades 1 & 2)

Place paper, pencils, rulers, yardsticks, string, scissors, and a measuring tape at the supply center. Then ask, "Is the perimeter of your hand greater than, less than, or equal to the perimeter of your foot?" Allow the children to figure out how to measure and how to compare. Don't rush to sweep in and help them out.

What you hope each child will do is this: Trace her hand (fingers together, not spread) and foot onto paper, then cut out each tracing. Carefully wrap the string around the outer edge of each piece, cutting the string in each case so the ends meet at the starting point. Finally, measure the string and compare lengths.

If they don't get it after some thought, begin some prompting. Don't give it all to them, though. Let them do it. They can!

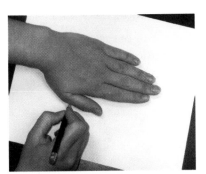

Step 1: *Trace the hand and foot.*

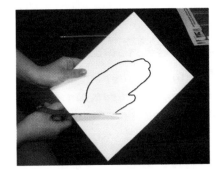

Step 2: *Cut out the tracings.*

Step 3: *Wrap the string around the circumference of each, and cut the string.*

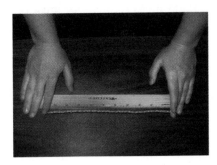

Step 4: *Measure each length of string.*

⬡ Handy References

Kindergarten concentrates on nonstandard measurement, first grade makes the transition to standard, and second grade works on standard measure. It's our job to help our students develop benchmarks for specific units so they really do know that their dads are not 8 feet tall and their baby sisters do not weigh 3 ounces!

To put all these abstractions in a "real-world" context, help students understand that the distance between their knuckles is about an inch, 4 sticks of butter is a pound, the tip of their baby finger is close to a centimeter wide, and a paper clip weighs about 3 grams. When you are in class and read that the baby rhinoceros is 5 feet tall, demonstrate that height. When a foot, a meter, or a centimeter comes up in conversation or in a book, show what a foot looks like, how tall a meter is, or how small a centimeter is.

To build benchmarks for liquid measures, label juice and milk cartons with their capacities. Use labels like, "This container holds 1 gallon" and "This milk carton holds ½ pint."

Use colored water to fill the kinds of plastic bottles your students commonly see. Label the bottles with sentences like, "This bottle holds 1 liter" and "This bottle holds 2 liters."

Place metric-sized bottles on one shelf in the classroom and standard pint, quart, and gallon bottles on another shelf. Seeing these bottles helps children develop benchmarks.

Fill those plastic bottle with colored water. These still need to be labeled, so students can use them as benchmarks for liquid measures.

WHY METRICS MATTER

In most of the world the metric system is the measure of choice. The U.S. is one of the very few countries that still use English customary measurement for most transactions of daily life. At the grocery store, we use pounds to measure the produce we buy. When we watch football games, we measure success in yards.

However, *scientific* measurement in the U.S. is metric. Doctors write prescriptions in metrics, and if you break a bone, the physician describes the break in centimeters. NASA uses centimeters to design rockets and to report findings. So in your science lessons, *stick to metrics.* Record in metrics the growth of the bean plants, the length of the worms, the weight of the class hamster, or the distance the Matchbox car traveled.

During your math instruction, alternate between metric and English customary so your students will become fluent in both. However, do not combine metric and standard in one lesson. That's a sure recipe for confusion!

⬡ Find Benchmarks

If you want your students to really know 10 inches, ask them to find 10 things that are 10 inches. This repetition helps them understand that measure.

⬡ Stringbean Small

Introduce your children to "Stringbean Small," a basketball player from a Jack Prelutsky poem of the same name, included in his book *The New Kid on the Block*. In the poem Stringbean is 8' 4" tall. Let the children help you measure, sketch, and then paint a mural of a basketball player who is exactly 8' 4" tall—just like Stringbean.

Begin by rolling out approximately 9 feet of bulletin board paper or butcher paper. Ask kids to predict how tall 8' 4" is and to mark their predictions in pencil on the paper; then measure and find who was closest. Mark Stringbean's measurements on the mural. Add a line at the side that measures his height in 1-foot increments. Then place the mural in a prominent space where it can stay all year long. Stringbean will become your standard point of reference. When you read that the baby giraffe was 8 feet tall, point out that there's a place in your room where 8 feet is indicated! When you read how tall the home-team sports heroes are, you need only to look to your very own Stringbean to get a sense of their height.

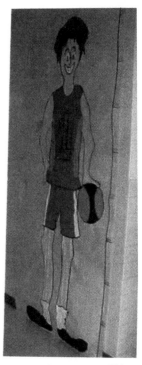

Stringbean Small is 8' 4" tall.

ART TIP: Let the class decide the color of Stringbean's hair, uniform, shoes, and socks, and pick the number he should be wearing on his uniform. Lightly sketch these in. (If you do too much, it takes away from the kid art.) Then let them paint. The final touch is a thick black outline around the character, giving definition to the wobbly edged paint that you're bound to end up with when kids paint. Select a child to draw that final outline with black marker.

Variation: For use in alternating years, I created "Daisy Divine, who was 7 feet 9. From her hat to the floor she was 8 feet 4." If you want a female represented, maybe your own version of Daisy will grace your halls!

Daisy Divine is 7' 9". From her hat to the floor, she's 8' 4"—just like Stringbean!

STANDARD

Money is not listed in the NCTM standards, but it is a part of many states' measurement standards.

EXPECTATION: Many states expect students to recognize coins, to understand the value of coins, to total coins to a dollar, and to count change.

Get Help from Home

This is an area that can really use help from home. Ask parents to help their children count change each evening. When students count change they are counting by 1, 5, 10, and 25, so they are also practicing addition and some multiplication. Five nickels is 5 x 5 = 25! Encourage parents to let children "pay the bill" and count change when they shop.

The Moneyed Class

Any of these classroom activities will help reinforce basic money concepts for your students.

Future financiers have to start somewhere. Let them sort plastic coins.

- Pour out the plastic coins and let children sort them. This takes close observation.

- Let children make rubbings of plastic or real coins so they can learn to recognize different denominations and to see differences and similarities.

- Give children magnifying glasses and let them study the coins.

- Have them use plastic coins to make patterns.

- Bring in grocery-store circulars from newspapers and ask children to count the necessary change for some of the items listed on a page. Explain that once students have counted out the change, they're to leave that money on the ad for the item they're "purchasing."

- At a center, place newspaper circulars advertising small items for less than $5.00. Also set out money stamps, an ink pad, and paper. Let each child cut out an item, glue it to the paper, and stamp the value of that item on the paper.

- Start with blank dice—one die for every 4 students—and write a coin value on each face of each die. (Since you'll be working with only 4 coin values—1¢, 5¢, 10¢, and 25¢—you'll have to repeat 2 coin values each time.) Divide students into groups of 4; then pass out the coin dice and plastic money to the groups. Within each group, children take turns rolling the die. Each time a child rolls, he takes a coin that equals the amount on the die. If a child accumulates 5 pennies, he turns them in for a nickel, and so on. The first child to reach $1.00 is the winner.

- A variation on this game is to give each child $1.00 in change before play begins. In this case, each roll indicates how much you *take away,* and the first person to hit zero is the winner.

- Begin with a dollar in change. Swap coins as you read the poem "Smart," from *Where the Sidewalk Ends,* by Shel Silverstein. Ask, "Was he *really* smart?"

- Move plastic money or real coins into a clear jar as you read *The Penny Pot,* by Stuart J. Murphy. Encourage children to talk about the total in the pot each time a coin is added.

- Give each of several small groups $1.00 in plastic coins. Place your own dollar's worth of plastic coins on the overhead projector. With the class, read *Alexander, Who Used to Be Rich Last Sunday,* by Judith Viorst. Together, remove the appropriate coins as Alexander spends his money.

Students choose items in a grocery circular and "pay" for them with plastic coins.

Data Analysis & Probability

Tehyi Hsieh said, "The schools of the country are its future in miniature." As adults, we look at and analyze data daily. We want our students to understand how to read that data and how to judge it for accuracy. If I want to know what flavor ice cream is most popular among 6-year-olds, should I consult a group of 6-year-olds or should I interview a sixth-grade class? Of course the answer is simple; why would you go to anyone other than 6-year-olds to find out the preferences of 6-year-olds?

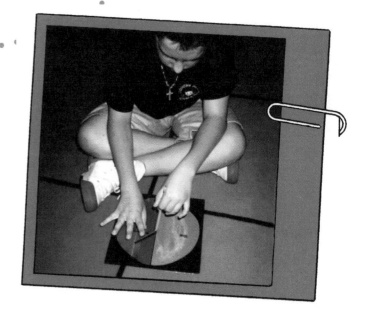

Let's take it a step further. If I want to find out who serves the best pizza in my town, where should I gather data: from the town next door, from a survey conducted by one pizza restaurant's customers, or from a survey included in the town newspaper that reaches all citizens of the town? Our students need to be able to answer that question, picking the newspaper survey because it reaches many people and because those people are representative of the database in question.

If our students are future consumers of news, they're also future consumers of products, which means that they must be able to interpret the "data-based" claims of advertisers, too. We're always reading or hearing claims about "The Most Popular," "The Top," "98% of Dentists Surveyed," or "Voted Best." What we don't always know is who was asked and how many were in the survey. If there were only 40 dentists in the survey and they all worked for the Brite Smile Toothpaste Company, is the claim that "98% think Brite Smile toothpaste is the best" really valid?

Whether data appears in the news columns or in advertising, it's often presented in the form of graphs. Pharmaceutical companies, political campaigns, stock analysts, and the newspaper sports page all present information in graphs. Often such a visual format can convey certain kinds of information more clearly and succinctly than sentences can. The consumers and voters of tomorrow are in our classrooms today, and they need to be able to deal with information communicated in this way.

In other words, we want our students to become savvy consumers of information. Let's make sure they learn to ask intelligent questions when they conduct surveys, that they graph with care, and that they know how to interpret data when it's presented in varied forms.

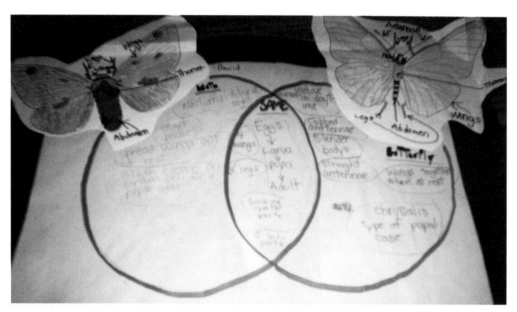

Children learn to understand data when they have a chance to manipulate data. Venn diagrams like this one let them apply their math skills to other areas of the curriculum.

STANDARD

Formulate questions that can be addressed with data and collect, organize, and display relevant data to answer them.

EXPECTATION: Pose questions and gather data about themselves and their surroundings.

◆ Collect Questions & Answers

Encourage children to come up with the question for a class survey and to phrase the question clearly. Praise children when they say, "I wonder which kind of cookie is this class's favorite." That is a perfect springboard into a kid-generated graph! Then prod: "How can we determine the answer to that?"

Once questions are established, children will need to decide how to collect the information. Allow your class to collect their information their way. If they have an unreasonable method, they might not see its impracticality until they try it themselves. Most likely their survey will flop and the children will realize their method had some fatal errors. Live and learn.

WHATEVER WAY WORKS

As far back as 1825, Warren Colburn wrote that children were successful in mathematics when they were allowed to pursue their own methods rather than being given rules (Copley 1999). I'm thrilled when children tell me how they solved the problems I have posed and doubly thrilled when other children arrive at the same answer using different methods.

Everyone's Favorite Subject

Children love questions about themselves. After all, they are their own favorite subjects! Before creating actual graphs, set the stage by modeling the kinds of data-oriented questions you want students to be asking—basing your questions on the children themselves. You can use the questions listed here as prompts for your students' own questions or for graphs that you create. Certainly this hefty list is not intended for you to use all at once. Use these topics sparingly and you'll be set for the entire school year!

- Are you left- or right-handed?

- What color is your hair?

- What color are your eyes?

- Is your hair curly or straight?

- How many are in your family?

- Are you the oldest, youngest, or only child in your family?

- What color car does your family drive the most?

- Did you travel over summer vacation?

- When is your bedtime? (This can be a real eye-opener for the teacher.)

- What is your favorite flavor ice cream?

- What is your favorite cereal?

- What is your favorite type of pizza?

- What is your favorite candy?

- What is your favorite dessert?

- What is your favorite lunch from the cafeteria?

- What is your favorite summer movie?

- What is your favorite (insert author name here) book?

- Which do you like more: a shower or a bath?

- What is your favorite kind of cookie?

- What is your favorite part of Thanksgiving dinner?

- What is your favorite tradition at holiday time?

- What is your favorite holiday treat?

- What is your favorite restaurant?

- What do you like best at McDonald's/Burger King/Wendy's/Taco Bell?

- What color socks are you wearing today?

- What is your favorite jelly-bean color?

- What is your favorite sports team?

- Do you like ice cream or Popsicles better?

- What is your favorite board game?

- Have you ever gone camping?

- Are you wearing a collar today or not?

~~~~~~~~~~~~~~~~~~~~~~~~~~~

EXPECTATION: Sort and classify objects according to their attributes and organize data about the objects.

~~~~~~~~~~~~~~~~~~~~~~~~~~~

◆ Sort & Classify Objects

See pages 61–64 for lessons and activities that give children practice in sorting and classifying.

◆ Data & Objects

When all the data have been gathered, the children need to learn how to organize it. Should all cheese-pizza pictures be put in a vertical row and all pepperoni-pizza pictures in a horizontal row? Of course not. That's why our students need to hear us think out loud as we move pieces or objects.

Maybe each child has brought the box from his favorite cereal to class (a fun way to graph, but one that takes up lots of space). Then you would say things like, "How shall we sort these boxes?" Pause. Long pause. "I can see we have sweetened cereal and unsweetened cereal. So maybe I could make a row of sweetened cereal. I'll line up all the boxes from sweetened cereal in a row." Move the sweetened-cereal boxes into one line.

Next say, "Now I'm going to take the unsweetened cereals and put *those* boxes in line. I need to be sure to put my first box in line with the first sweetened-cereal box. I have to make sure that each box in my unsweetened line is directly across from a box in the sweetened line, especially since the boxes are all different sizes." Demonstrate how the boxes line up, one across from the other.

Say, "I can see that the line of sweetened cereal is much longer. There are 9 more boxes in the sweetened-cereal line than in the unsweetened-cereal line. I can tell because 9 boxes in the sweetened-cereal line don't have matches."

A child places a paper triangle that represents his favorite kind of pizza. At the same time, he helps to build a class graph.

EXPECTATION: Represent data using concrete objects, pictures, and graphs.

◆ Concrete Objects

Younger children will need many more experiences with concrete objects than second graders will, but all children do better when their understanding is grounded in the concrete.

Use milk cartons from the cafeteria as manipulatives to determine whether more students prefer chocolate milk or white milk. Ask your students to bring their empty (*totally* empty!) milk cartons back to class after lunch. Before the students return to class, lay out 2 signs on the floor: "We like chocolate milk" and "We like white milk."

When they return to the classroom, ask the children to go to the sorting area one at a time, placing their cartons in a line on the floor. Each child should place her carton above the sign that describes her favorite type of milk. As the milk cartons are positioned, ask questions like, "What do you think will be the class favorite?" "What does the trend look like? Can you predict what will be the most popular?" Remind children to line up the cartons using one-to-one correspondence so that you can easily judge which type is more popular.

Of course, talk, talk, talk about the graph: "What is the difference between the number of those who like chocolate milk and the number of those who like white milk?" "What do you think we can say about all first graders?" "Can we predict what most students in our school would prefer?"

A QUICK TIP

The one-to-one method really helps children see that one set has more than the other. Lining up items to show one-to-one correspondence is also great practice for children when they count independently.

Chocolate milk is the most popular. Now you can milk your graph for all it's worth.

GREAT GRAPHING MATERIALS

In addition to milk cartons, any of these real-world materials make great classroom graphs.

- Use toothpaste boxes brought from home to determine the class's favorite brand of toothpaste.

- Use boxes from TV dinners for favorite TV dinner.

- Use shoes to show whether students' footwear has Velcro or not (or show white soles vs. not white soles or white laces vs. dark laces).

- Use real cookies for favorite cookies.

- Use Beanie Babies to indicate reptiles, mammals, and so on or to compare numbers of animals living on land against numbers of animals living in water.

- Use hats to show whose hats are related to sports teams and whose are not.

Once students have experience with the concrete, bump things up a notch by representing data with pattern blocks or blocks from the block center. These materials are still concrete, of course, but because they're more abstract than the milk cartons and shoes, they help to build student understanding.

STANDARD Select and use appropriate statistical methods to analyze data.

EXPECTATION: Describe parts of the data and the set of data as a whole to determine what the data show.

◆ Comparing Data

When you or the children come up with a data question, ask, "How can we best display our information?"

In the early stages of gathering data, young children can form a line and use one-to-one correspondence to determine which set has more and which set has less. For instance, all children who prefer chocolate milk form a straight line, one behind the other, with their hands on their hips. All children who prefer white milk also form a straight line with their hands on their hips, positioning themselves next to the chocolate-milk line. The first child in the chocolate-milk line touches elbows with

Line kids up hand-to-shoulder. That's a memorable lesson in one-to-one correspondence.

the first child in the white-milk line. This continues until one child does not have someone to stand elbow-to-elbow with.

That's one way to know which set has more. "I can see that there are more children who like chocolate milk than white milk because this line is longer and because there are several children in the chocolate-milk line who have no one to stand elbow-to-elbow with."

Once the children have been lined up in rows for a human graph, you can add variety by demonstrating one-to-one data correspondence in other ways, too. Have the children reach out their arms and stand hand-to-shoulder. Have them sit on the floor and line up foot-to-foot. Or have them raise their arms and link hands "London Bridge" style. Any of these shows one-to-one correspondence.

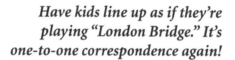

Have kids line up as if they're playing "London Bridge." It's one-to-one correspondence again!

Here's one more "kid graph." As soon as one line runs out of kids to pair off, it's clear that the other line has more.

◆ Venn Diagrams

Venn diagrams are a fun way to collect and record information. Your students already know that from their algebra activities (see pages 63–64). If there are 2 questions, there are 2 circles. If there are 3 questions, there are 3 circles. The circles should intersect/overlap as shown in the illustration.

Create simple cut-outs and challenge students to find the right spots for their "shirts." This is a great way to get students engaged in learning.

Remind your little Einsteins that for anything listed where the circles overlap, both attributes exist. If there is a picture outside both or all circles, then that item is in the universal set.

ART TIP: Circles for Venn diagrams can be tough to draw. To make circles, create a primitive compass. Cut string that is a bit longer than the radius of the circle you want to draw. Tie the string around a pencil, near the pencil's tip. Hold the other end of the string down with one hand on the chart or bulletin-board paper. With the other hand, swing the pencil point around to form a circle.

VENN POSSIBILITIES

Any of these ideas can be presented in picture form with a Venn diagram.

- **Do you have brothers? Do you have sisters?**

- **Does your family have Christmas traditions? Does your family have Hanukkah traditions?**

- **Do you like basketball? Do you like baseball?**

- **On Valentine's Day: Did you wear red? Did you wear pink? Did you wear white?**

- **Does your name have 2 word parts? Does your name begin with a J? Does your name have an a?**

- **What is your hair color? Is your hair curly? Are your eyes brown?**

Venn diagrams do not have to be circles! Make hearts for Valentine's Day, animal shapes when comparing animals, insect shapes when comparing insects.

◆ Reading Connections

Venn diagrams can also explain information. They can compare 2 stories, such as *The Little Red Hen* and *The Little Yellow Chicken,* and they can show what students have learned from research, too.

When first and second graders present information they've researched, it doesn't always need to be in the form of a written report. Assign each reading group or pair of children 2 similar topics. Let them research the topics and then present their new knowledge in Venn diagram form.

There are countless topics that can be compared and contrasted in Venn diagrams, but these work especially well.

- Crocodiles and alligators
- Squirrels and chipmunks
- Rabbits and kangaroos
- Dolphins and porpoises
- Frogs and toads
- Rats and mice
- Owls and bats
- Butterflies and moths

- Foxes and wolves
- Helen Keller and Thomas Edison
- George Washington and Abraham Lincoln
- 2 or 3 folktales
- Different versions of the "Cinderella" story from around the world
- 2 or 3 Native American tribes

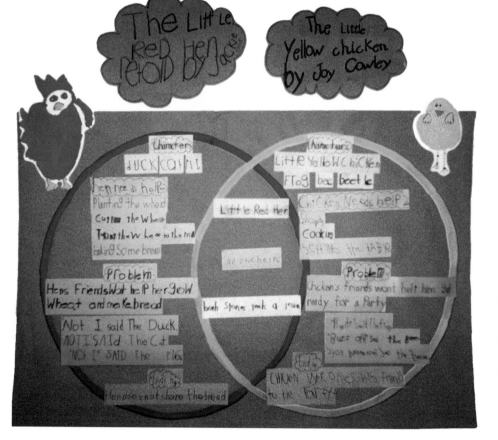

Students enjoy using Venn diagrams to show what they've learned. This one compares the Little Red Hen with the Little Yellow Chicken.

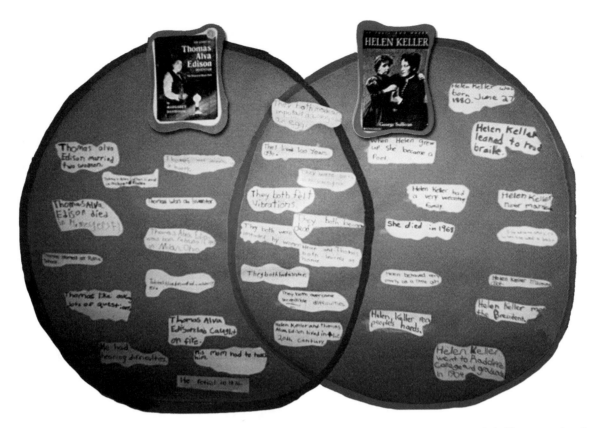

Remember to use Venn diagrams to help your students see the similarities and differences in the topics you are studying across the curriculum. When you want students to show what they've learned, Venn diagrams can be a great alternative to research papers. This one, created by one of my reading groups, compares Thomas Edison and Helen Keller.

GRAPH EVERY WHICH WAY

Your students should see a variety of graphs, including everything from circle graphs to both vertical and horizontal bar graphs. And don't forget the pictographs!

◆ 2-Dimensional Graphs

To assemble a graph, children should identify parts as they add information. You can guide this production by following steps something like these:

- First ask, "What shall we call our graph? Every graph must have a title, so what will our title be?"

- Once everyone agrees on a title, write that title at the top of the paper or board where your graph will go.

- Ask, "Do you want to make this a vertical graph or a horizontal graph?" Allow them to decide.

- If they want to make a vertical graph, say, "All graphs must have labels for the columns. If we make this a vertical graph, the labels go on the x-axis. You tell me what labels we should write."

- After labeling the columns add, "The y-axis is on the left side." Older children should label that axis with numbers.

Of course, if the students choose a horizontal graph, you say, "If we make this a horizontal graph, our labels will be on the y-axis. You tell me what labels to write."

After labeling the columns add, "In this case the numbers are shown on the *x*-axis, which is at the bottom of the graph."

If this is a pictograph, say, "Every pictograph must have a key, so we need to add a key. A key explains what one picture means." Ask a child to make a small drawing that you will glue to the side of the graph. For example, a pizza picture might represent one child's favorite kind of pizza.

◆ Circle Graphs

This is an easy way to make a circle graph that shows the colors of M&M candies in a mini-bag. (You can also do this with Froot Loops or Trix cereal.) After spilling out the candies from the bag, the child sorts them by color. Then she draws a circle large enough so that all the candies can fit when lined up along the circumference of that one circle. (For an easy way to draw the circle, see page 133.) She puts all red candies together on the edge of the circle, then does the same with each other color, keeping the space between candies consistent. Finally, starting at each point where a new color begins, she draws a line from the outside of the circle to the center. She's made a pie chart!

The student sorts the candies, lines them up on the perimeter of the circle, and draws a line from the start of each new color to the center. It's a pie chart!

STANDARD Develop and evaluate inferences and predictions that are based on data.

EXPECTATION: Discuss events related to students' experiences as likely or unlikely.

◆ Likely, Unlikely & Impossible

Before class, start with 3 short sentence strips. Write the word *likely* on one, *unlikely* on the second, and *impossible* on the third. Then take 9 longer sentence strips. On each, write one of the sentences in the box on the next page (or substitute your own).

In class, pass out the "likely," "unlikely," and "impossible" strips to 3 children and ask them to stand in front of the class with them. Review the 3 terms, asking students to explain the meaning of each one. Be sure kids understand.

Next, pass out the longer sentence strips, giving one to each of 9 students. One child at a time stands in front of the class and reads her sentence. As a group, students decide where that child should stand. Should she stand behind the "likely," the "unlikely," or the "impossible" strip?

After all sentence strips have been read, the class generates other likely, unlikely, and impossible sentences. If possible, tie this in with the topics you are studying in science and social studies.

Finally, each child selects a theme and writes about it, being sure to include a statement that's likely, a second that's unlikely, and a third that's impossible. For example: "Today I am going to walk my dog when I get home. It's likely my little dog will bark at every big dog we see. It's unlikely we'll be able to walk one mile because my dog has short little legs. It's impossible my dog will decide to drive the car around the block instead of walking."

Continue using this language as you read aloud to children. "Is it likely that Patricia Polacco will pass her keeping quilt on to her daughter?" "Is it likely that Strega Nona will continue to make pasta and that Big Anthony will stop eavesdropping?" "Is it likely Jane Yolen will take her children owling?"

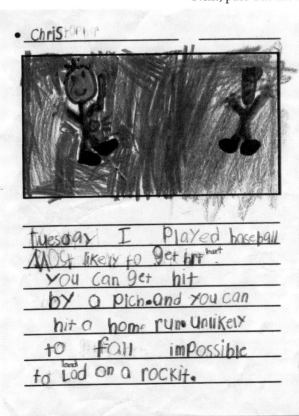

This baseball player understands "likely," "unlikely," and "impossible."

FOR EXAMPLE

These are the kinds of sentences you might want to use for sorting into likely, unlikely, and impossible categories.

- The cafeteria will be busy on pizza day and on chicken-nugget day.
- We will read many books this year.
- Penguins will waddle on ice.
- The cafeteria ladies will serve jelly-bean salad and chocolate soup tomorrow.

- The governor will teach reading groups today.
- We'll have recess for one hour today.
- Abe Lincoln will read his autobiography.
- Babe Ruth will coach a Little-League team.
- Our next field trip will be to the planet Mars.

◆ Noticing the Trend

Let's say you're graphing a subject like "What is your favorite part of the Thanksgiving meal?" Before you begin the graph, remind the children that this is not a vote. There is not a winner. This is a collection of information. Ask each child to predict what the most children will say and to write or draw his prediction before you gather the information.

"What's your favorite holiday tradition?"

Then pass out an index card to each student, and ask each child to draw her favorite part of the meal. As you hand out the cards, say, "Remember to hold your card in a wide rectangle. Now draw whatever is your favorite part of Thanksgiving dinner."

Once drawings are complete, call one child at a time to place her card on the floor so that, taken together, all the cards make a graph. To do this the child has to think about what her card shows and where to place it. "Hmm. I'm the first person to like sweet potatoes. Where should I put my card?"

After 5 or 6 cards are on the floor, say, "Notice the trend. We have 3 people who like stuffing best, 2 who like pie, and 1 who likes cranberry sauce. Now that we have this information, would any of you like to make a prediction about the outcome of our graph?"

After a few kids have expressed their opinions, continue calling children to add their information to the graph, stopping a few more times to notice the trend and checking to see if the trend is changing.

Once the graph is complete, of course it needs to be interpreted. This is a very serious part of graphing that is often overlooked. You need to model for students how to interpret the information on their graph, guiding them through the answers to questions like these:

● How many people were surveyed? *We have data from 24 kids.*

● What can you discover when you compare the data in each column? *More people like pumpkin pie than cranberry sauce and more people like drumsticks than pumpkin pie. Hardly anyone likes sweet potatoes. More than half of our class likes drumsticks best.*

A QUICK TIP

When children begin comparing data, it's easier for them to start with 2 attributes than with several. Compare numbers of pumpkin-pie lovers and drumstick lovers before you add cranberry sauce to the mix!

● Can you make a prediction onto a larger population? *If we asked other children in our grade this question, drumsticks might be their favorite or their second favorite. Most kids would like sweet potatoes least.*

You can write or type these interpretations as the children dictate their observations about the graph. Once the children have done this together, they can handle the interpretation of other graphs on their own.

ART TIP: When you're using pictures to compare data, it's important that all the pictures are the same size. One way to manage this is to hand out blank 3 x 5-inch index cards and say to students, "Hold this in a wide rectangle and draw on it." (You can also have them glue cut-out pictures to the cards.) Just try to make sure everybody orients the cards the same way. If someone draws with the index card oriented the other way, you have no choice but to turn the drawing on its side. Otherwise the results are not clear.

 STANDARD Understand and apply basic concepts of probability.

EXPECTATION: There are no specific expectations given for this standard.

◆ Likely & Unlikely Reading

Begin this study with the outrageously bizarre *And to Think That I Saw It on Mulberry Street,* by Dr. Seuss. While reading this story, ask questions using the terms "likely" and "unlikely." Is it likely that a boy would walk home from school? Is it likely he'd talk to his dad when he got home? Is it likely that on his walk he'd see a kid on a bike or a lady with a baby buggy? What else is likely on a walk home? When you get to the absurd ask, "Is it likely he saw zebras?" and so on.

Read *Probably Pistachio,* by Stuart J. Murphy, and emphasize the probable occurrences in the daily life of this boy. Unlike *Mulberry Street,* this book focuses on realistic probability, so it provides a great balance.

The Prediction Bag

A little bit of mystery doesn't hurt when you're trying to develop prediction skills. Just follow these steps.

- Before class, place 8 green and 2 yellow crayons in a gift bag.

- In class, shake the bag and say, "What do you think I have in this bag?"

- Accept predictions. Praise students' thoughts.

- "Is it likely that I have toasters in here?" Pause for laughter.

- "Is it possible that I have hamsters in here?" Pause for more laughter.

- Let a child reach in and pull out 1 crayon, show it to the class, and then return it to the bag.

- Say, "Okay, so what is one sure thing you can say about this bag?"

- You hope that someone will say, "There's 1 crayon in the bag."

- Let another child pull out a crayon and ask, "Could this be the same crayon? How do we know?"

- After a few pulls and discussion say, "Interesting! We have pulled 5 times. We pulled out 4 green crayons and 1 yellow crayon. Can anyone make a prediction about what is in this bag?"

- "Could there be any red crayons or orange crayons in the bag?"

- Do not be tempted to tell them the answers or explain the logic involved in any of these questions. They have to construct this knowledge through their own experience.

- Ask questions like, "Do you think there are any purple crayons in here?" Insist that students explain their answers.

- Say, "I'll tell you that we do have 10 crayons in this bag. We know that some are green and some are yellow. We don't know if there are any other colors. Can you predict how many I have of each?"

- "Is it likely there are 8 purple crayons?"

- "Is it unlikely there are 7 blue crayons?"

- Let each child in the class reach in and pull out 1 crayon, then return it to the bag. The larger the sampling of crayons, the more accurate the prediction will be.

A little mystery helps to keep students engaged in predicting.

● When everyone has had a turn drawing from the bag and making a prediction, pull out the crayons one at a time in front of the class. Discuss how the results were predictable because of the results of each pull.

After many experiences the children will begin to understand and believe that if there are 8 crayons of the same color they can never be certain that each time they pull out a green crayon it is the same green crayon. If your children have a difficult time with this concept, you may want to put a tally mark on the crayon each time it is pulled out. While this will distract from the prediction of colors of crayons in the bag, it may help drive home the concept that we cannot be certain that the same crayon was pulled each time.

This activity can be repeated many times. Use Hershey's Kisses and Hugs, mini-candy bars, socks of different colors, links, Unifix cubes, markers, checkers, crayons, bingo chips, or blocks. Whatever you choose, make sure every piece that goes into the bag at the same time has the same shape and the same feel. (My little sharpies figured out that a Baby Ruth feels different from a Snickers. Watch out for those junior detectives in your class who will "feel" their way around that bag before making predictions!)

As you repeat this game, vary the number combinations as well as the types of objects being counted. Maybe a bag will have 1 black crayon and 9 red ones, another bag 5 black socks and 5 red socks, and yet another 2 black dice and 8 white ones. The more you play this game the better your children's predictions and use of mathematical language will be.

A Quick Tip

I like to use 10 items in the bag since 10 is an important benchmark for kids and an easy number when it comes to talking fractions and percentages with the older first graders and with second graders. For some children it will be appropriate to say, "Look at this bag. There are 8 blue socks and 2 white socks. This bag is 8 of 10 or 80% blue and 2 of 10 or 20% white." Don't shy away from giving your kids language like this that they can really sink their teeth into!

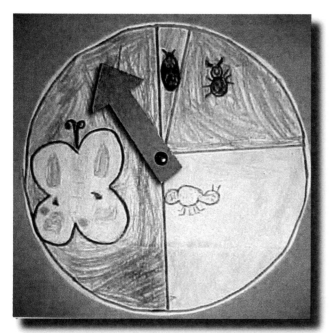

If you don't want to use a paper clip, you can put the "spin" in a spinner with a brad and a construction-paper arrow.

◆ Probability Spinners

Game spinners are great tools for helping children to see two things: first, there is a greater chance of landing on the space that is largest; and second, if all spaces are the same size, there is an equal chance of landing on each space.

Collect spinners from old board games and make some from scratch in several different sizes. (Teacher-made spinners representing seasons and holidays are always popular.) Include some spinners that have equal-size spaces and some that don't. To put the "spin" in a spinner, put a paper clip in the center of the spinner, place the point of a pencil through the open space in the paper clip and touching the board, and then spin the paper clip.

After whole-group instruction let your students play and explore with the spinners on their own at tables or centers. Be sure to follow up this exploration with discussion.

◆ Make-Your-Own Spinners

After the students understand the probabilities associated with different configurations of spinners, it's time for them to make their own spinners to show that understanding. The spinners they make should demonstrate which outcomes are the most and the least likely. Ask students to create "themed spinners." For instance, a sports-themed spinner might have hockey sticks, baseball bats, basketballs, and/or footballs. A weather spinner could have clouds, sunshine, raindrops, and/or snowflakes.

Use wrapping paper to decorate the bases of homemade spinners.

INSTRUCTIONS FOR MATH JOURNALS

"Describe the spinner you made, using mathematical language."

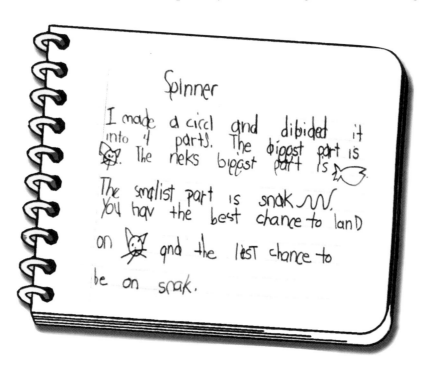

POTENTIAL SPINNER QUESTIONS

Ask students to answer some of these questions in their math journals.

- What did you predict about your spinner? What were the outcomes of your spins? What conclusions did you come to?

- What part of the spinner did you land on most often? Is this what you expected? Why?

- What part of the spinner did you land on *least* often? Is this what you expected? Why?

- What if you spun another 10, 15, 50 times? Do you think the results would be the same?

- What if your friends had the same spinner? Would their results be similar to your results? What makes you think that?

- Think about the spinners in the board games you've played. Are all spinners fair? Or do some spinners have a better chance of landing on one section than on others?

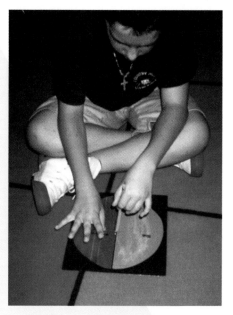

A pencil and a paper clip work with a homemade spinner.

REPRODUCIBLES, PATTERNS & RESOURCES

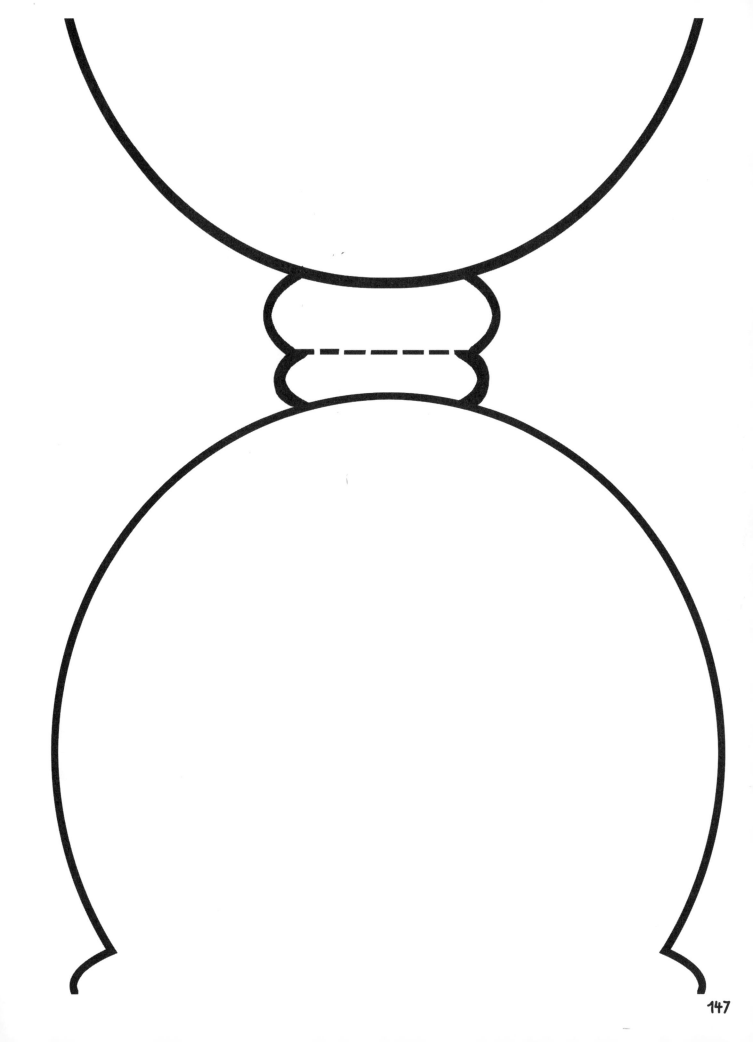

fold

(cut 2)

half of pillow

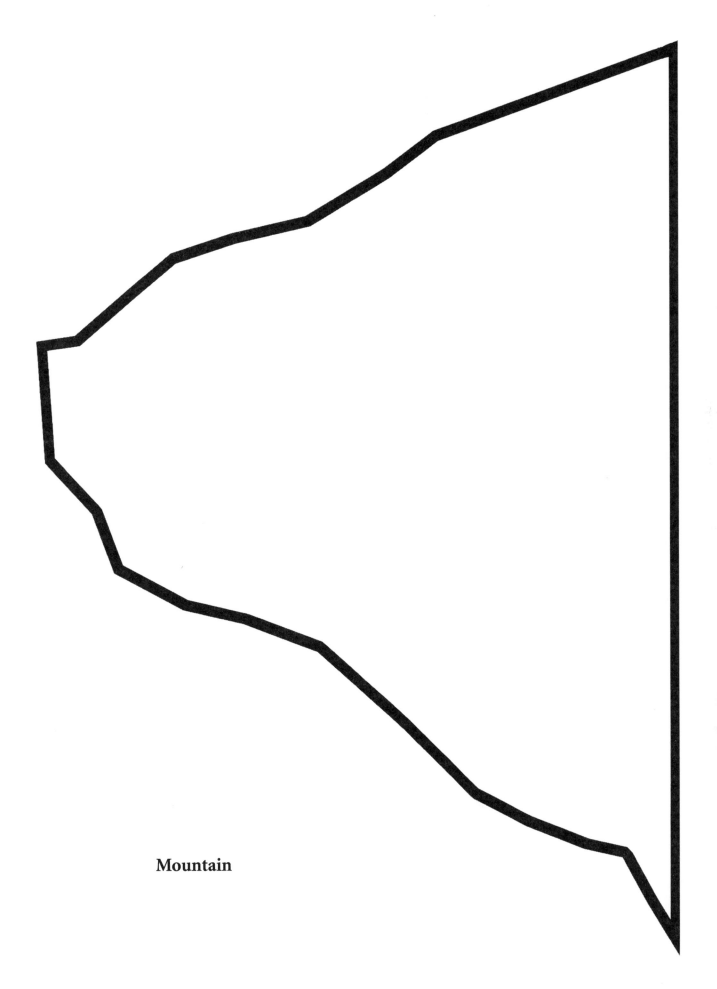

Mountain

Lava

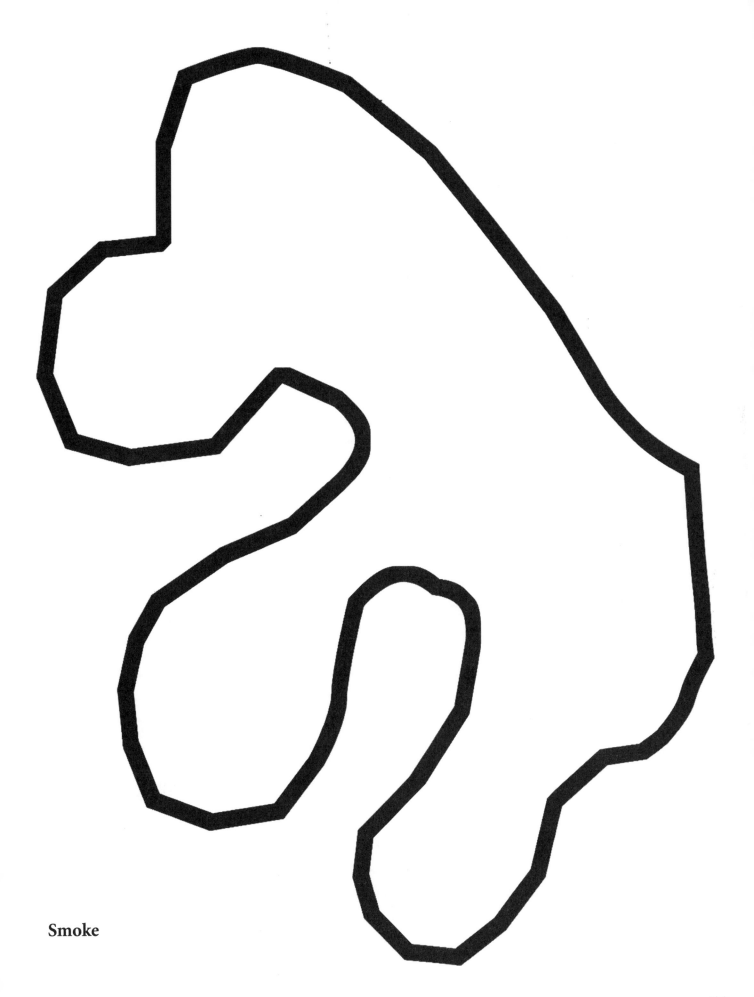

Smoke

155

100 CHART GAME BOARD

1	2	3	4	5	6	7	8	9	10
11	12	13	14	15	16	17	18	19	20
21	22	23	24	25	26	27	28	29	30
31	32	33	34	35	36	37	38	39	40
41	42	43	44	45	46	47	48	49	50
51	52	53	54	55	56	57	58	59	60
61	62	63	64	65	66	67	68	69	70
71	72	73	74	75	76	77	78	79	80
81	82	83	84	85	86	87	88	89	90
91	92	93	94	95	96	97	98	99	100

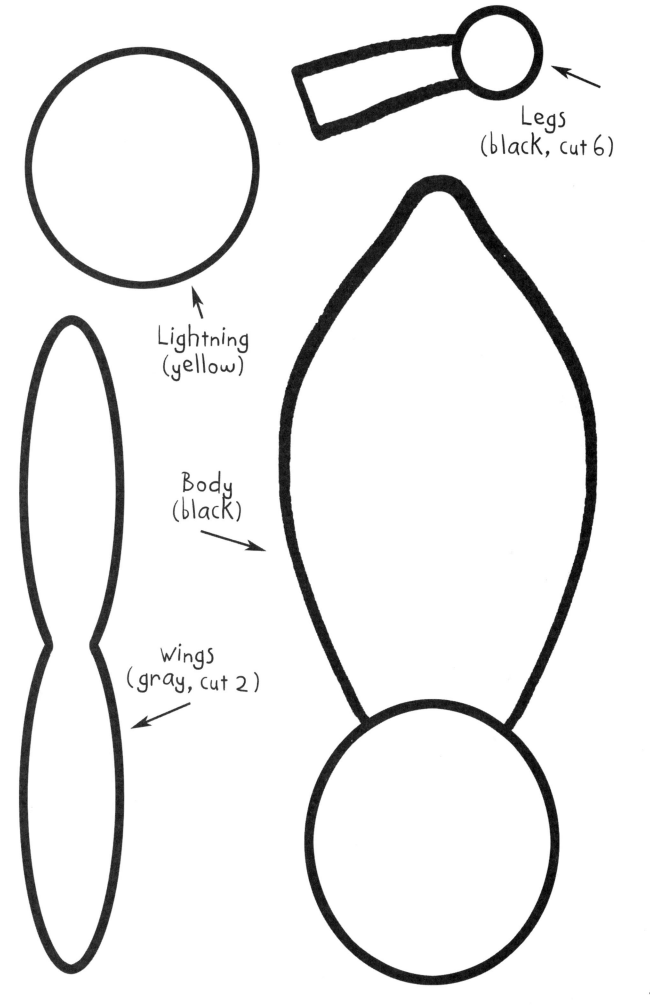

Legs
(black, cut 6)

Lightning
(yellow)

Body
(black)

Wings
(gray, cut 2)

157

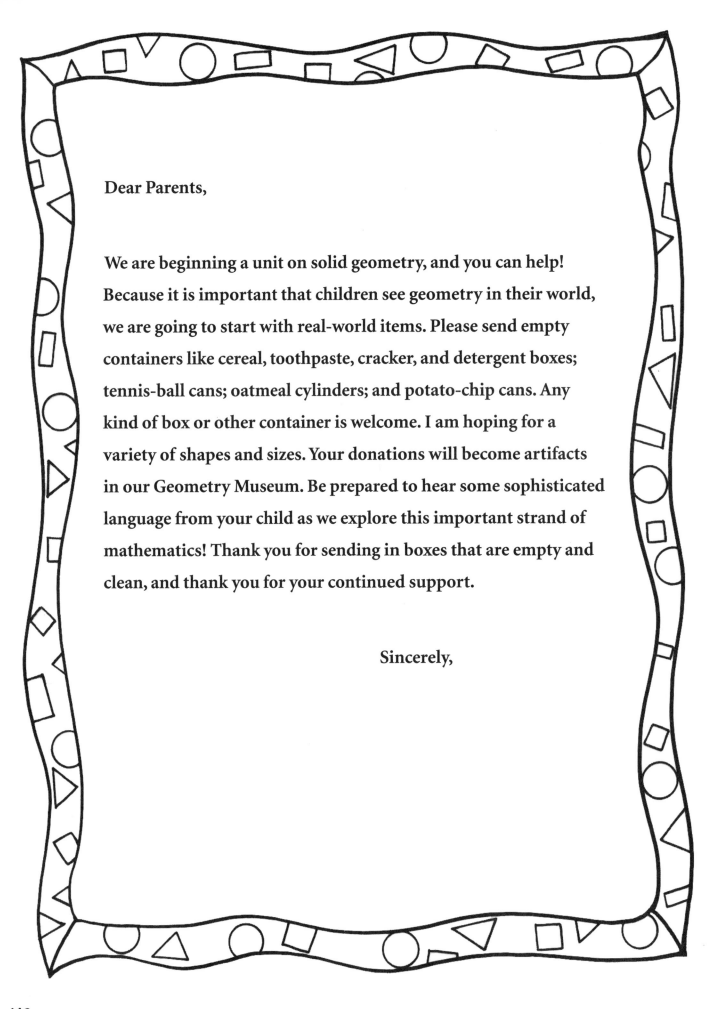

Dear Parents,

We are beginning a unit on solid geometry, and you can help! Because it is important that children see geometry in their world, we are going to start with real-world items. Please send empty containers like cereal, toothpaste, cracker, and detergent boxes; tennis-ball cans; oatmeal cylinders; and potato-chip cans. Any kind of box or other container is welcome. I am hoping for a variety of shapes and sizes. Your donations will become artifacts in our Geometry Museum. Be prepared to hear some sophisticated language from your child as we explore this important strand of mathematics! Thank you for sending in boxes that are empty and clean, and thank you for your continued support.

Sincerely,

PIGOMETRY

Oink! Oink! Make a pig following these directions:

1. Choose black, pink, or gray paper.
2. Cut 7 similar circles for the head, eyes, body, snout, and nostrils.
3. Cut 2 congruent triangles for ears.
4. Cut 4 congruent rectangles for legs.
5. Glue all pieces together to assemble one fabulous pig.
6. Twirl a pipe cleaner for a tail.
7. Write 4 sentences in your math journal describing the geometry in your pig.

GEOME-TURKEY

Make a turkey following these directions:

1. Choose tan, gray, or brown paper for your turkey's body, plus other colors for the feet, comb, and beak.

2. Cut 4 similar circles for the head, body, and eyes.

3. Cut 2 congruent rectangles for legs.

4. Cut 4 congruent rectangles OR 4 congruent triangles for feathers.

5. Cut 2 congruent triangles for feet.

6. Cut a triangle for the comb.

7. Cut a rhombus for a beak.

8. Glue all pieces together to create one handsome turkey. When you get to the beak, fold the rhombus so it looks like a triangle, with half glued down and half sticking up.

9. In your math journal, write 4 sentences describing the geometry in your turkey.

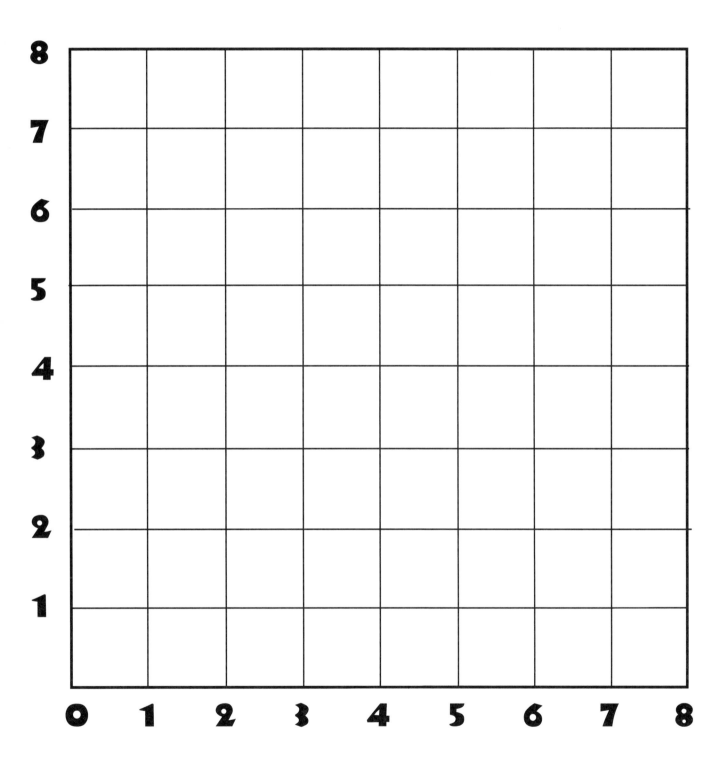

GEOMETRY HUNT BY

_____

See how many of these shapes you can find in your home. You can draw a picture of what you find, or you can write each object's name, or both.

I found these shapes in my home:

_____

Spheres

_____

Rectangular prisms

_____

Cubes

_____

Triangular prisms

_____

Pyramids

_____

Cylinders

_____

Animal	Length or Height in _____
Yellow jacket	
Stag beetle	
Praying mantis	
Sparrow	
Lobster	
Skunk	
Boa constrictor	
Hyena	
Gorilla	
Rhinoceros	
Elephant	
Whale	

Children's Literature with Math Connections

Some of the books listed here are recommended for use with the activities in this book. The rest are included simply because they offer wonderful ways to teach math through children's literature.

● Number & Operations

Anansi the Spider, by Gerald McDermott

Arctic Fives Arrive, by Elinor Pinczes

Bear Snores On, by Karma Wilson

Betcha!, by Stuart J. Murphy

The Doorbell Rang, by Pat Hutchins

Each Orange Had 8 Slices, by Paul Giganti, Jr.

Eating Fractions, by Bruce McMillan

Elevator Magic, by Stuart J. Murphy

Every Buddy Counts, by Stuart J. Murphy

Gator Pie, by Louise Mathews

Give Me Half!, by Stuart J. Murphy

The Grapes of Math, by Greg Tang

Henry the Fourth, by Stuart J. Murphy

Math for All Seasons, by Greg Tang

Miss Spider's Tea Party, by David Kirk

Missing Mittens, by Stuart J. Murphy

My Little Sister Ate One Hare, by Bill Grossman

The Napping House, by Audrey Wood

One Grain of Rice, a folktale illustrated by Demi

100 Hungry Ants, by Elinor Pinczes

A Remainder of One, by Elinor Pinczes

Roar! A Noisy Counting Book, by Pamela Duncan Edwards

Sea Squares, by Joy N. Hulme

Seaweed Soup, by Stuart J. Murphy

The Tale of Tom Kitten, by Beatrix Potter

Ten Flashing Fireflies, by Philemon Sturges

Ten Go Tango, by Arthur Dorros

Ten Little Mice, by Joyce Dunbar

Too Many Kangaroo Things to Do!, by Stuart J. Murphy

Traveling to Tondo, by Verna Aardema

12 Ways to Get to 11, by Eve Merriam

Two of Everything, by Lily Toy Hong

Two Ways to Count to Ten, by Ruby Dee

The Water Hole, by Graeme Base

You might also like to share the poem "Nine Mice" from *The New Kid on the Block,* by Jack Prelutsky.

Algebra

Beep Beep, Vroom Vroom!, by Stuart J. Murphy

The Button Box, by Margarette S. Reid

Double the Ducks, by Stuart J. Murphy

A String of Beads, by Margarette S. Reid

3 Little Firefighters, by Stuart J. Murphy

Two of Everything, by Lily Toy Hong

Geometry

Captain Invincible and the Space Shapes, by Stuart J. Murphy

Circus Shapes, by Stuart J. Murphy

A Cloak for the Dreamer, by Aileen Friedman

Coyote Steals the Blanket: A Ute Tale, retold by Janet Stevens

The Fly on the Ceiling, by Dr. Julie Glass

Grandfather Tang's Story, by Ann Tompert

The Greedy Triangle, by Marilyn Burns

My First Book of Shapes, by Diane Namm

Shape Race in Outer Space, by Calvin Irons

The Village of Round and Square Houses, by Ann Grifalconi

The Warlord's Puzzle, by Virginia Walton Pilegard

Wayne's New Shape, by Calvin Irons

You might also like to share the poem "Shapes" from Shel Silverstein's *A Light in the Attic.*

⬡ Measurement

Alexander, Who Used to Be Rich Last Sunday, by Judith Viorst

The Best Bug Parade, by Stuart J. Murphy

Clocks and More Clocks, by Pat Hutchins

Get Up and Go!, by Stuart J. Murphy

The Grouchy Ladybug, by Eric Carle

Henry Hikes to Fitchburg, by D.B. Johnson

Inch by Inch, by Leo Lionni

The Kissing Hand, by Audrey Penn

Mr. Archimedes' Bath, by Pamela Allen

The Penny Pot, by Stuart J. Murphy

Room for Ripley, by Stuart J. Murphy

The Very Hungry Caterpillar, by Eric Carle

Who Sank the Boat?, by Pamela Allen

You might also like to share the poems "Smart" from *Where the Sidewalk Ends*, by Shel Silverstein, and "Stringbean Small" from *The New Kid on the Block*, by Jack Prelutsky.

◆ Data Analysis & Probability

And to Think That I Saw It on Mulberry Street, by Dr. Seuss

The Best Vacation Ever, by Stuart J. Murphy

The Button Box, by Margarette S. Reid

Chrysanthemum, by Kevin Henkes

Probably Pistachio, by Stuart J. Murphy

Tiger Math: Learning to Graph from a Baby Tiger, by Ann Whitehead Nagda

Tikki Tikki Tembo, retold by Arlene Mosel

REFERENCES

Brooks, Jacqueline Grennon, and Martin G. Brooks. 1993. *In search of understanding: The case for constructivist classrooms.* Alexandria, VA: Association for Supervision and Curriculum Development.

Burns, Marilyn. 2000. *About teaching mathematics: A K–8 resource.* Sausalito, CA: Math Solutions Publications.

Chambers, Donald L., ed. 2002. *Putting research into practice in the elementary grades.* Reston, VA: NCTM.

Copley, Juanita V., ed. 1999. *Mathematics in the early years.* Reston, VA.: NAEYC and NCTM.

Ferrini-Mundy, Joan, Glenda Lappan, and Elizabeth Phillips. 1997. Experiences with patterning. *Teaching Children Mathematics,* February: 282–88.

Ginsburg, Herbert P., and Sylvia Opper. 1998. 3rd ed. *Piaget's theory of intellectual development.* Upper Saddle River, NJ: Prentice Hall.

Hiebert, James, et al. 1997. *Making sense: Teaching and learning mathematics with understanding.* Portsmouth, NH: Heinemann.

Kilpatrick, Jeremy, Jane Swafford, and Bradford Findell, eds. 2001. *Adding it up: Helping children learn mathematics.* Washington, DC: National Academy Press.

National Council of Teachers of Mathematics. 2000. *Principles and standards for school mathematics.* Reston, VA: NCTM.

Resnick, Mitchel, and Fred Martin. 1998. *Digital manipulatives: New toys to think with.* Cambridge, MA: MIT Media Laboratory. http://el.www.media.mit.edu/groups/el/papers/mres/chi-98/digital-manip.html.

Van de Walle, John A. 2003. *Elementary and middle school mathematics: Teaching developmentally.* Boston, MA: Allyn & Bacon.

Wadsworth, Barry J. 1989. *Piaget's theory of cognitive and affective development.* White Plains, NY: Longman.

Whitin, Phyllis, and David J. Whitin. 2000. *Math is language too: Talking and writing in the mathematics classroom.* Urbana, IL: National Council of Teachers of English.

INDEX

Note: Page numbers in italics indicate reproducibles.